Astronomy Media Workbook

Fifth Edition

for

The Cosmic Perspective
The Essential Cosmic Perspective

Bennett Donahue Schneider Voit

Michael C. LoPresto
Henry Ford Community College

San Francisco Boston New York
Cape Town Hong Kong London Madrid Mexico City
Montreal Munich Paris Singapore Sydney Tokyo Toronto

Vice President, Editorial Director: Adam Black, PhD
Senior Acquisitions Editor: Lothlórien Homet
Media Producer: Deb Greco
Assistant Editor: Ashley Anderson
Managing Editor, Production: Corinne Benson
Production Supervisor: Liz Winer
Marketing Director: Christy Lawrence
Manufacturing Buyer: Michael Early
Production Services: GGS Book Services
Main Text Cover Design: Jeff Puda
Supplement Cover Design: Seventeenth Street Studios
Manufacturing Manager: Pam Augspurger

Cover Credit: The Earth view on the cover was created by ARC Science Simulations using ARC's Face of the Earth computer model of a planet. New Milky Way image (top right corner) is courtesy of Dr. Mark A Garlick.

MasteringAstronomy Tutorials
Authored and developed by the following team:
Content Director and Lead Author: James Dove, Metropolitan State College of Denver
Educator Team Contributors: Nicholas Schneider, University of Colorado, Boulder
Fran Bagenal, University of Colorado, Boulder
Erica Ellingson, University of Colorado, Boulder
Kevin McLin, University of Colorado, Boulder

Media Developer:
Managing Director: David Hegarty
Cadre design Pty, Ltd.
http://www.cadre.com.au

Multimedia Director:
Senior Programmer: Chris Kemmett
Interactive Programmer: Bill Stern
Graphic Designer: Janet Saunders
Graphic Designer: Melanie Halliwell
Animator: Leigh Russell

ISBN 0-8053-9593-8

1 2 3 4 5 6 7 8—B&B—09 08 07 06

www.aw-bc.com

ACKNOWLEDGEMENTS

Voyager: SkyGazer, College Edition CD ROM™

Developed by Carina Software based on the Voyager suite of products.

If you enjoyed using *Voyager: SkyGazer*, College Edition, v3.7, and would like to explore the heavens in greater depth, Carina Software offers astronomy software packages tailored to your level of interest. You may download demo versions from their website. Carina Software offers the following:

Voyager, v3.78

This program is the most powerful and easy-to-use planetarium software available. With a new streamlined interface, you can create realistic sky simulations in multiple windows. Using the latest celestial data and deep space images, display the complex motions of planets, comets and asteroids. Show the rich star fields of the Milky Way containing hundreds of thousands of stars. Voyager is both a powerful learning resource and an indispensable tool for exploring the heavens. With the release of Voyager, v3.78, SkyPilot is now included as a standard feature. SkyPilot allows you to control the position and motion of a telescope.

Voyager: SkyGazer, v3.6

For more movies and tutorials, you may wish to upgrade to the complete SkyGazer package. It is the ideal companion for backyard astronomers first exploring the wonders of the night sky. It is a celestial adventure for all ages.

Carina Software
602 Morninghome Road
Danville, CA 94526
www.carinasoft.com
Carinasoft@aol.com
Tel: 925-838-0695
Fax: 925-838-0535

SYSTEM REQUIREMENTS

System Requirements of Carina Software's Voyager: SkyGazer, College Edition for use with the Astronomy Media Workbook

Windows: 366 MHz Pentium, Windows 98/2000/XP, 64 MB RAM minimum.

Macintosh: 233 MHz, OS 9.2, 10.2, 10.3, 64 MB RAM minimum.

Both platforms: 12 MB hard disk space, 1024 × 768 screen resolution, thousands of colors, CD ROM drive, Apple's QuickTime 4.1, 5.0, or 6.0 and Adobe Acrobat Reader. If you do not have QuickTime or Acrobat Reader on your system, you may obtain these free plug-ins at www.apple.com/quicktime and www.adobe.com.

INSTALLATION

This program runs right off the CD, but you may copy it to your hard drive if you wish.

Windows: Insert the CD-ROM, open Windows Explorer and find your CD drive, and click on the SkyGazer icon to launch the program.

Macintosh: Insert the CD-ROM and double-click on the SkyGazer icon to launch the program.

DOCUMENTATION

A user guide for Voyager, on which SkyGazer is based, is provided on its CD-ROM in the Help Menu. You'll need Adobe Acrobat to read these files. For a free Adobe Acrobat plug-in, visit www.adobe.com.

System Requirements for MasteringAstronomy (www.masteringastronomy.com)

Windows: 250 MHz CPU; OS Windows 98, NT, 2000, XP

Macintosh: 233 MHz CPU; OS 9.2, 10.2, 10.3; Red Hat Linux 8.0, 9.0

All: 64 MB RAM, 1024 × 768 screen resolution. Browsers (OS dependent): Firefox 1.0, Internet Explorer 6.0, Mozilla 1.7, Netscape 7.2, Safari 1.3. Flash 7.0. System requirements are subject to change. See website for current requirements.

MasteringAstronomy™ is powered by MyCyberTutor by Effective Educational Technologies.

Technical Support:

www.aw-bc.com/techsupport
Monday–Friday, 9 AM–6 PM Eastern time (U.S. and Canada)

Contents

Introduction

TO THE INSTRUCTOR

MasteringAstronomy

The MasteringAstronomy website, www.masteringastronomy.com, is designed to assist students taking an introductory astronomy course using either *The Cosmic Perspective* or *The Essential Cosmic Perspective*. Your students receive free access to the MasteringAstronomy website with each copy of a new textbook. Students who buy used textbooks can buy access to the site at www.masteringastronomy.com.

This first half of this Media Workbook is written specifically to accompany the 22 Mastering Astronomy Web Tutorials. Within each activity there are questions to accompany each tutorial lessons that are meant to consolidate student understanding of the main concepts in the tutorial. A laboratory activity follows the questions. This activity makes use of one of the many interactive tools in the tutorial for a more in-depth investigation of a single important concept from the tutorial.

The questions and lab activities can be used in various ways when accompanying either *The Cosmic Perspective* or *The Essential Cosmic Perspective*. The Tutorials can be used as homework assignments in

MasteringAstronomy Tutorial	Tutorial Title	The Cosmic Perspective Chapter	The Essential Cosmic Perspective Chapter
1	Scale of the Universe	1	1
2	The Seasons	2, S1	2
3	Eclipses	2	2
4	Phases of the Moon	2	2
5	Orbits and Kepler's Laws	3, 4, 7	3, 4
6	Motion and Gravity	4	4
7	Energy	4	4
8	Light and Spectroscopy	5	5
9	The Doppler Effect	5	5
10	Telescopes	6	5
11	Formation of the Solar System	8, 9, 11, 12	6
12	Shaping Planetary Surfaces	9	7
13	Surface Temperature of Terrestrial Planets	10	7
14	Detecting Extrasolar Planets	13, 24	6

(continued)

MasteringAstronomy Tutorial	Tutorial Title	The Cosmic Perspective Chapter	The Essential Cosmic Perspective Chapter
15	The Sun	13, 24	10
16	Measuring Cosmic Distances	15, 20	11
17	The Hertzsprung-Russell Diagram	15	11
18	Stellar Evolution	15, 17 ,18	12, 13
19	Black Holes	18, 19, 21	13
20	Detecting Dark Matter in Spiral Galaxies	19, 22	16
21	Hubble's Law	20, 23	15, 17
22	Fate of the Universe	22	16

a large lecture class or as in-class activities for smaller classes if a computer lab is available. The table above correlates the MasteringAstronomy Web Tutorials with the chapters of both textbooks.

Since the MasteringAstronomy Web Tutorials provide activities on the full spectrum astronomical topics, they are also ideal for use in the laboratory portion of an introductory astronomy course or in a separate introductory laboratory course. Having students complete and submit the lesson questions and laboratory activity for one tutorial per week makes an ideal curriculum for a classroom laboratory course, and the independent nature of the tutorials makes them ideal for an independent or online class.

To aid in managing an independent or online laboratory course, the MasteringAstronomy website includes a feature that allows instructors to create a class that students must join. Once they have joined they will be able to submit their work to the instructor, which will provide the instructor with a complete record of the work the students have done in the tutorial. The questions and activities from the Media Workbook can then be submitted by the students for a grade. They can either be mailed or dropped off at an assigned location, a department office, the instructor's office, a mailbox, etc. The availability of 22 tutorials provides flexibility in which topics to cover in a typical 15-week semester.

Voyager: SkyGazer

SkyGazer is the educational version of the interactive desktop planetarium program Voyager by Carina Software. All students also receive a multi-platform SkyGazer CD with each new copy of *The Cosmic Perspective* or *The Essential Cosmic Perspective*.

Each activity in the SkyGazer portion of the Media Workbook is designed for students to work through simulations of various astronomical events. In each activity the student is given detailed directions on how to carry out a simulation and is instructed to report their results on a separate Results Sheet that can be submitted to the instructor.

As an alternative to submitting these results sheets, each SkyGazer activity also includes 10 multiple choice questions that are designed to test student understanding of each activity's important concepts. There is also an "open-ended" activity where the student is asked to carry out an investigation related to the activity *without* explicit instructions. This allows evaluation of not only understanding of the astronomical concept, but also of how much facility with the program has been learned. The table on the following page correlates the SkyGazer activities with the chapters of both textbooks.

Sky Gazer Activity	Activity Title	The Cosmic Perspective Chapters	The Essential Cosmic Perspective Chapters
1	Introducing *SkyGazer*	1	1
2	Motions of the Stars	2, S1	2
3	Celestial Sphere	2, S1	2
4	Motions of the Sun	2, S1	2
5	The Ecliptic	2, S1	2
6	Seasonal Constellations	2	2
7	The Seasons	2	2
8	Precession	2, 3	2
9	Proper Motion	2, 3	2
10	Phases of the Moon	2, 3	2
11	Solar Eclipse	2	2
12	Lunar Eclipse	2	2
13	The Inferior Planets	2, 3	3
14	The Superior Planets	2, 3	3
15	Observing the Planets	2, 3	3
16	Asteroids, Comets, and Meteors	12	9
17	Satellite and Spacecraft	24	18
18	The Solar System-Planets, Moons, and Rings	7–11	6–9
19	Stars and the HR Diagram	14–18	10–13
20	Galaxies—The Milky Way	19–21	14, 15
21	The Universe—Hubble's Law	22, 23	16, 17

Due to the observational nature of SkyGazer, the activities focus strongly on the motions of stars, the Sun and planets and less on the physical nature of objects like the MasteringAstronomy Web Tutorials. Although, as shown in the above table, Activities 18–21 are meant to accompany specific astronomical topics, one of SkyGazer's greatest strengths is that a compelling laboratory course can be built around it and the activities.

SkyGazer Activities 1–16 focus mostly on observations of the motions of the objects in the sky and can be used as weekly activities for a laboratory course focused on these subjects. The course could meet with weekly sessions in a computer-equipped laboratory classroom or as an independent-study course with the students completing activities and submitting the results on their own.

You can download an electronic version of the answer key for both the MasteringAstronomy Tutorials and the Voyager: SkyGazer Activities from the Instructor Resource Center online at http://www.aw-bc.com/irc.

TO THE STUDENT

With the *Astronomy Media Workbook*, MasteringAstronomy, and Voyager: SkyGazer, you are going to get even more out of your astronomy class by using a range of interactive and fun activities that will help you master the course's key concepts. The *Astronomy Media Workbook* accompanies both *The Cosmic Perspective* and *The Essential Cosmic Perspective*, and both textbooks include access to the MasteringAstronomy website and a CD of the program Voyager: SkyGazer.

MasteringAstronomy is a website designed to help you with the topics you will read about in the textbook and learn about in class. Your textbook already offers suggestions on which tutorials to use and how to use them to aid in the learning of important ideas in each chapter. This workbook provides additional questions and laboratory activities to help you master the concepts covered in each tutorial. If you bought a used copy of either *The Cosmic Perspective* or *The Essential Cosmic Perspective* that did not include a MasteringAstronomy access code, you can purchase access online at www.masteringastronomy.com.

Voyager: SkyGazer is an interactive desktop planetarium program. It allows you to set the sky to any place and time you wish to observe the motions of the stars or planets. Again, your textbook offers some suggestions on how to use SkyGazer. This workbook offers activities in which you will be instructed on how to simulate various astronomical events.

You may be using the MasteringAstronomy and SkyGazer with this workbook as a supplement to an introductory astronomy lecture class or as part of the main feature of an introductory astronomy laboratory course. Either way, the interactive nature of the software and the workbook should prove to be an invaluable resource in assisting your learning and enjoyment of introductory astronomy.

ACKNOWLEDGEMENTS

I thank Deb Greco, Ashley Taylor Anderson, and Liz Winer of Addison-Wesley for their efforts on this edition, and Adam Black and Claire Masson for providing me the opportunity to develop the workbook in the first place. I also thank the reviewers of previous editions, Stacie Kent, and colleague Steve Murrell for their valuable input that has helped make this workbook a better educational tool. Mostly, I thank the hundreds of students who took introductory astronomy laboratory at Henry Ford Community College. Without them to test all the ideas, this workbook would not have been possible.

Michael C. LoPresto
Dearborn, MI 2006

MasteringAstronomy Web Tutorials
LESSON QUESTIONS & LAB ACTIVITIES

MASTERINGASTRONOMY WEB TUTORIALS

1
Scale of the Universe

Your goals in this tutorial are to:

- understand the distances between astronomical objects relative to their sizes
- describe what a light-year and astronomical unit are
- identify how a light-year and astronomical unit compare to more familiar distances

LESSON 1

1. How far away is the Moon? How far is this in terms of Earth's diameter (see Figure 1-1)?

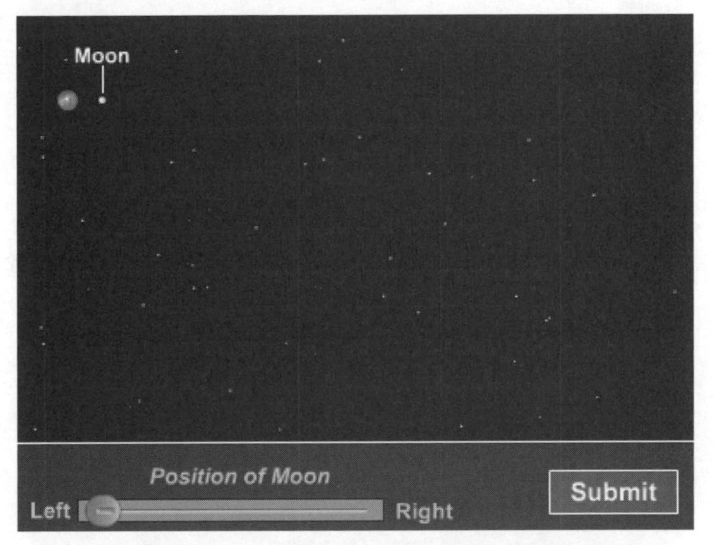

Figure 1-1 This tool from Lesson 1 determines the distance to the Moon using the same scale as the size of the Earth and the Moon.

2. What is the diameter of the Sun? How does it compare to the diameter of the Moon's orbit around the Earth? The diameter of the Earth?

3. How far is Jupiter from the Sun? How far is this in Astronomical Units (AU)? (The distance between the Earth and the Sun is 1 AU.)

4. How far is Pluto from the Sun in AU? How many times farther from the sun is Pluto than Jupiter?

LESSON 2

5. If the Sun were the size of a grapefruit, what would be the size of Jupiter? On this scale, how far from the Sun would Jupiter be? How far would Pluto be from the Sun on this scale? How far away would the Earth be?

6. If the Sun were the size of a grapefruit, the nearest star, Alpha Centauri, also would be about the size of a grapefruit. How far away from the Sun would Alpha Centauri be? On this scale, what location in the United States would correspond to this distance from Washington, D.C. (see Figure 1-2)?

7. Based on how long it takes light to travel from the Sun to the Earth, and the distance between the Sun and Earth in the model, determine the speed of light on the scale of the Solar System Scale Model on the Mall in Washington, D.C. (see Figure 1-2).

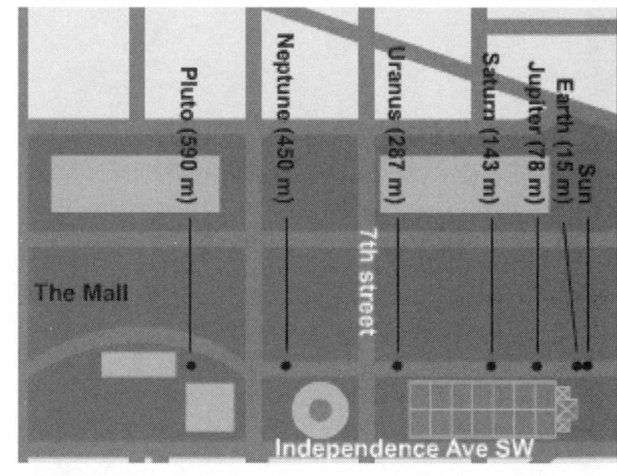

Figure 1-2 This solar system model on the National Mall in Washington, D.C., has a Sun the size of a grapefruit, which results in the planets being spread out over almost 600 meters—that is, over six football fields. Use this tool to help answer questions 6 and 7.

8. What is the distance between the Milky Way and Andromeda Galaxies? About how many times the Milky Way or Andromeda Galaxy's size is this distance?

LESSON 3

9. What is a light-year? How far is a light-year in meters? In AU? Write your answers in both regular and scientific notation. Use the tool shown in Figure 1-3 to help you convert from meters to AU.

10. The speed of light is about 3×10^8 meters per second. Write this number in regular notation. The speed of light can also be expressed as 186,000 miles per second. Write this number in scientific notation.

11. For each pair of objects in the table below, estimate the ratio of the distance between the objects to their size. Obtain the distances for your estimates from the order of magnitude tool shown in Figure 1-3. List the pairs from the smallest to the largest ratios.

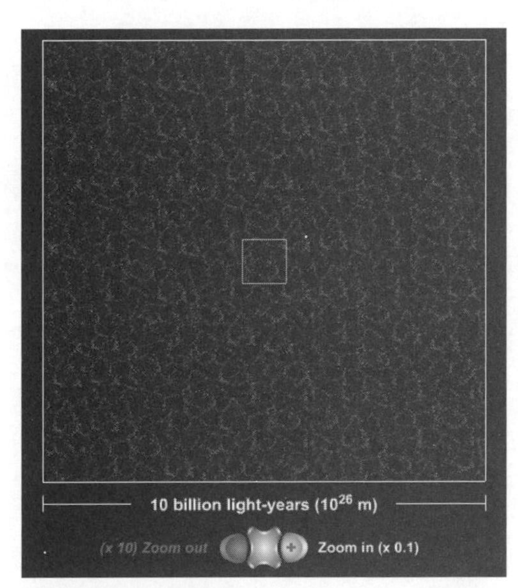

Figure 1-3 Shown here is the order of magnitude tool from Lesson 3. Use the distance given at the various orders of magnitude to make your estimates for the questions in Lesson 3.

Object	Approximate Size	Distance to	Distance/Size
Milky Way Galaxy	100,000 ly	Andromeda Galaxy =	
Sun	10^8 m	Alpha Centauri =	
Jupiter	10^7 m	Sun =	
Earth	10^6 m	Moon =	

12. Which distance to size ratio is the largest? The smallest? Which two of the distance-size ratios from question 11 are most similar in order of magnitude?

LAB ACTIVITY Scale of the Universe

The goal of this activity is to determine the time it takes at various speeds to travel the distance between astronomical objects.

Use the tool from Lesson 2 (shown in Figure 1-4) to answer these questions.

How long does light take to travel to the Earth from the Moon?

How long would it take a starship traveling at half the speed of light to get to Alpha Centauri?

The *New Horizons* spacecraft, launched in 2006, is expected to arrive at Pluto in about ten years. How fast is it traveling? Remember to use the tool in Figure 1–4!

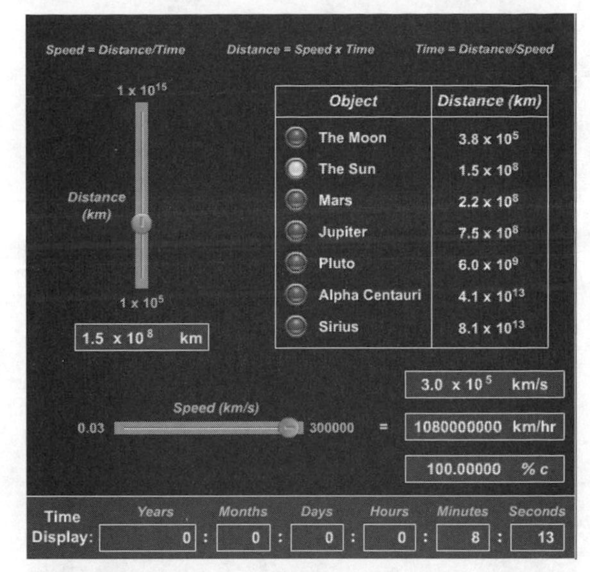

Figure 1-4 Use this tool from Lesson 2 to determine the time needed to travel different distances at different speeds.

The *Voyager* spacecraft, launched in 1977, travels at about 120 kilometers per second. How far away is it now? Is it still in the solar system? How long would it take *Voyager* to get to Sirius? Remember to use the tool in Figure 1–4!

2
The Seasons

Your goals in this tutorial are to be able to explain:

- the factors that affect the surface temperature of the Earth
- the reason why the Tropics are hotter than the Polar Regions
- the causes of the Earth's seasons, solstices, and equinoxes

LESSON 1

1. Which two factors affect the amount of sunlight received by a planet's surface?

2. If the Earth's axis were *not* tilted, what effect would this have on the seasons (see Figure 2-1)?

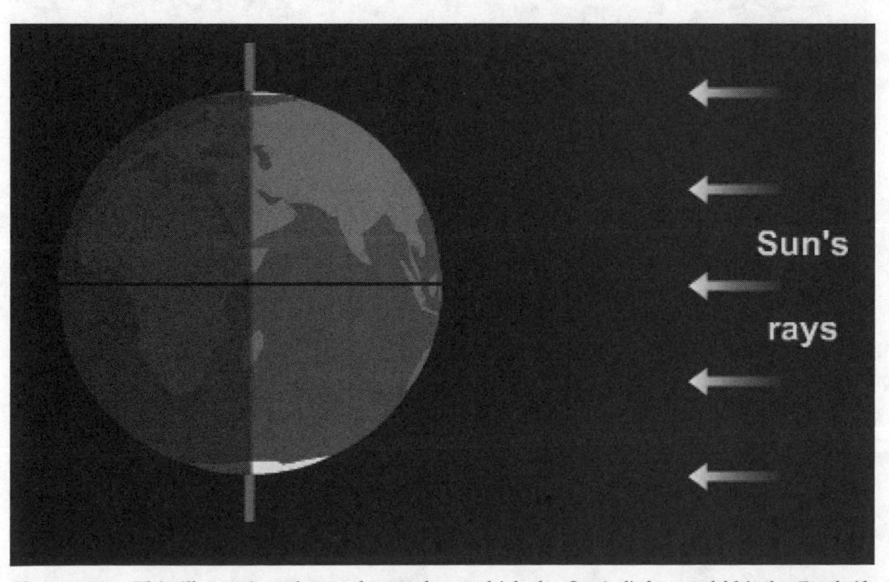

Figure 2-1 This illustration shows the angle at which the Sun's light would hit the Earth if the Earth's axis were *not* tilted.

3. When is the Earth closest to the Sun?

4. Based on your answer to question 3, does the variation in distance between the Earth and the Sun affect the seasons? Explain your answer.

5. Explain why the temperature at the equator is always hot and the temperature at the poles is always cold, despite the passage of the seasons.

LESSON 2

6. If it is summer in the United States, what season is it in Australia (see Figure 2-2)?

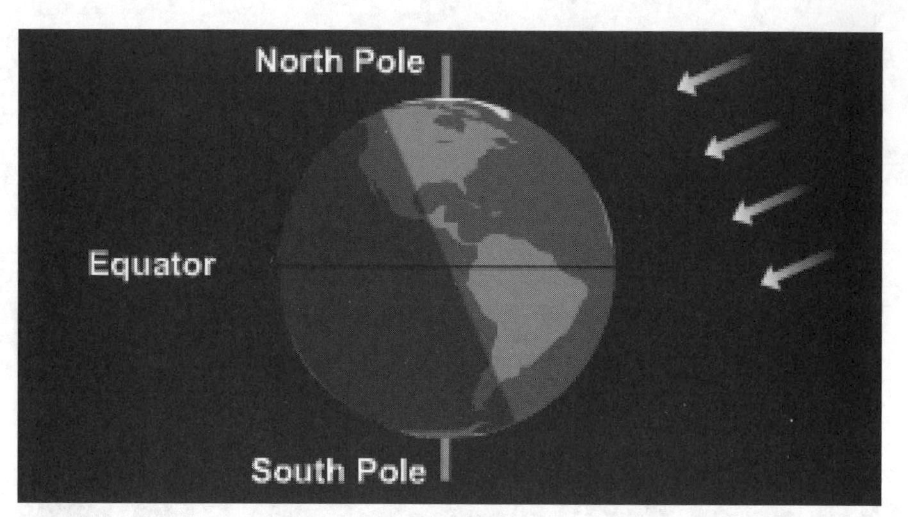

Figure 2-2 This diagram shows the angle at which sunlight hits the Earth on June 21.

7. Which regions of the Earth have the greatest seasonal changes as far as the amount of daylight and darkness they experience? Which have the least?

8. Where would you have to go in order to observe a midnight Sun? When would this occur?

9. If it is dark almost all day at a location in northern Alaska, what would be true of a location of similar southern latitude at the same time of year? You may have to try several of the slightly different versions of the tool in the lesson, shown in Figure 2-3, to answer the questions.

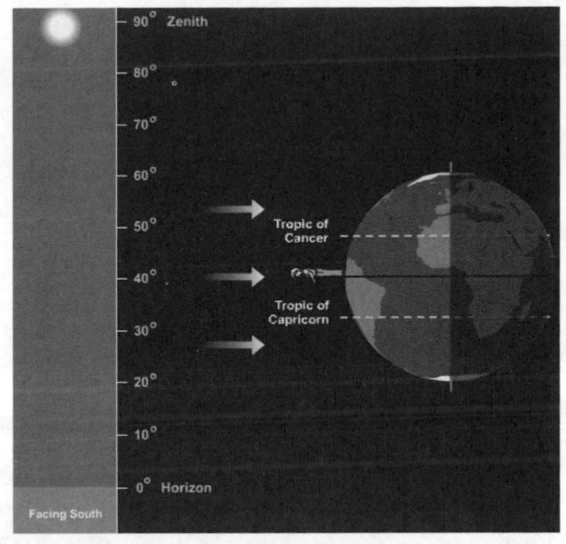

Figure 2-3 This tool shows how the Sun appears from different locations on Earth.

10. When is the Sun directly overhead at the equator?

11. Where is the Sun directly overhead on December 21, the winter solstice?

12. When is the Sun directly overhead at your location?

LAB ACTIVITY The Seasons

The purpose of this activity is to determine whether the Sun's altitude or duration (time the Sun is up) has more of an effect on temperatures at a given location.

Use the tool from Lesson 3 (shown in Figure 2-4).

Fill out the table for the person standing in Canada. You can estimate the hours the Sun will be up by looking at how much of the Earth is in daylight at the different latitudes. For example, at the Tropic of Cancer, about 2/3 of the Earth is in daylight, and 2/3 of 24 hours is 16 hours of daylight. Next, move the observer first to the North Pole and then to the Equator and fill out the table for each of these locations.

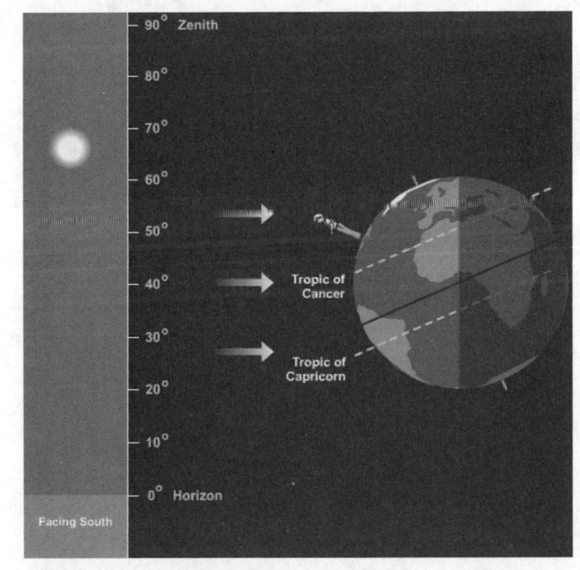

Figure 2-4 Use this tool from Lesson 3 to observe the Sun from different locations on the Earth. Be sure you use the version of the tool that is tilted exactly as shown above and allows you to move the person to different latitudes.

Location	Altitude of Sun	Hours Sun Will Be Up
North Pole		
Canada		
Equator		

Think of typical temperatures at the three locations you visited. Based on your data, can you determine which would be more important for warm weather, the Sun's altitude or how many hours it is up? Cite evidence from the above table for your answer.

3

Eclipses

Your goals in this tutorial are to:

- discuss the differences between solar and lunar eclipses
- explain why eclipses do not occur every month
- explain why eclipses are sometimes total and other times partial

LESSON 1

1. Explain what happens during a solar eclipse.

2. Explain what happens during a lunar eclipse.

3. Explain why eclipses do not occur every month (see Figure 3-1).

4. In which lunar phase do lunar eclipses occur? In which phase do solar eclipses occur?

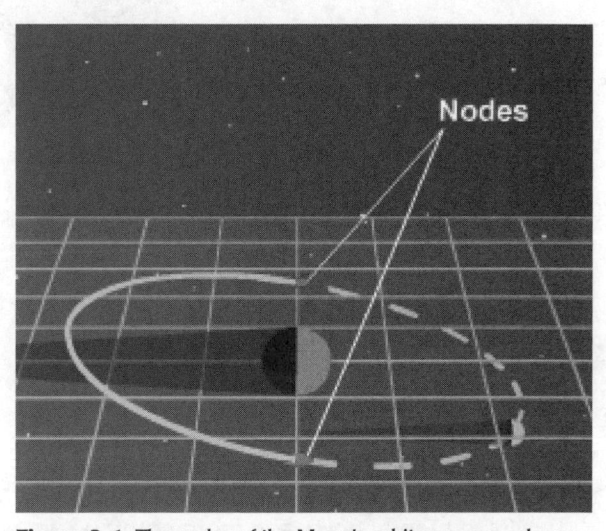

Figure 3-1 The nodes of the Moon's orbit appear as shown.

LESSON 2

5. What is an annular eclipse? Explain why they occur.

6. Explain why some observers of a solar eclipse see a total eclipse while others see only a partial eclipse. Also explain why some observers from Earth see no eclipse at all.

7. Is there one and only one specific place on the Earth from which a given solar eclipse can be seen? Explain your answer.

8. If a solar eclipse were being observed from the Earth, describe what an observer on the Moon would see when looking at the Earth (see Figure 3-2).

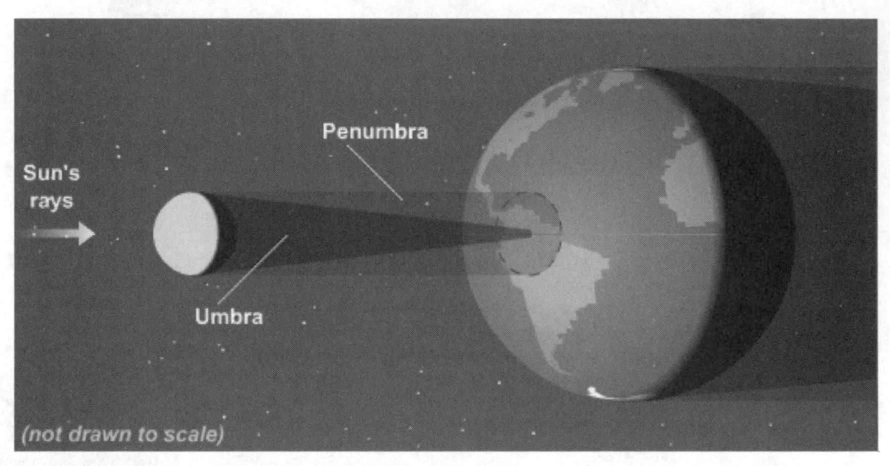

Figure 3-2 The Moon's shadow on the Earth creates an eclipse.

LESSON 3

9. From where on the Earth can a lunar eclipse be seen (see Figure 3-3)?

10. Which type of eclipse is more common, lunar or solar? Why?

11. Which type of eclipse do you think you are likely to observe more often in your lifetime? Explain your reasoning.

12. If a lunar eclipse were being observed from the Earth, what would an observer on the Moon see (see Figure 3-3)?

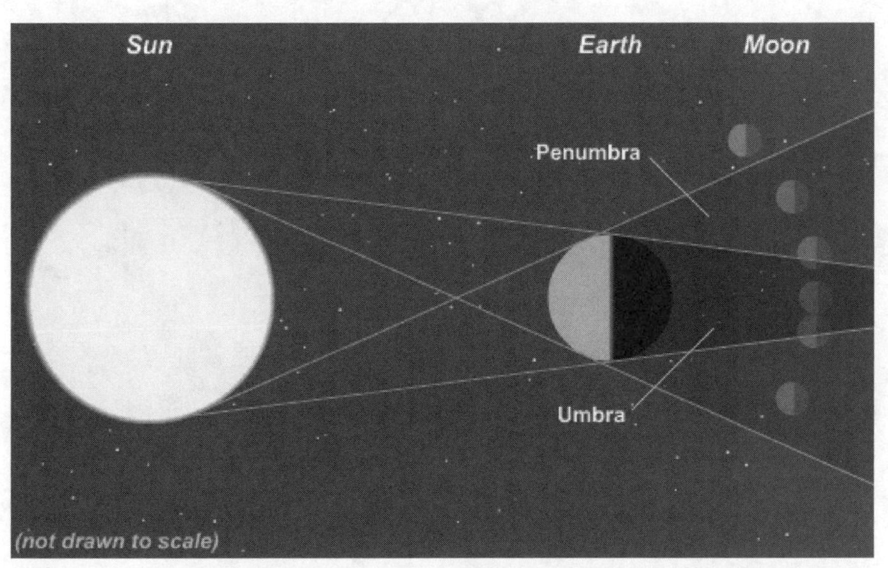

Figure 3-3 The orientation of the Sun, Earth, and Moon during a lunar eclipse appears as shown (see questions 9 and 12).

LAB ACTIVITY Eclipses

The purpose of this activity is to determine how differing distances between the Earth and the Moon will affect solar eclipses.

Use the tool from Lesson 2 (shown in Figure 3-4) to complete the table below.

The position to the right of the Moon in Figure 3-4 is position 1, and the position numbers progress counterclockwise.

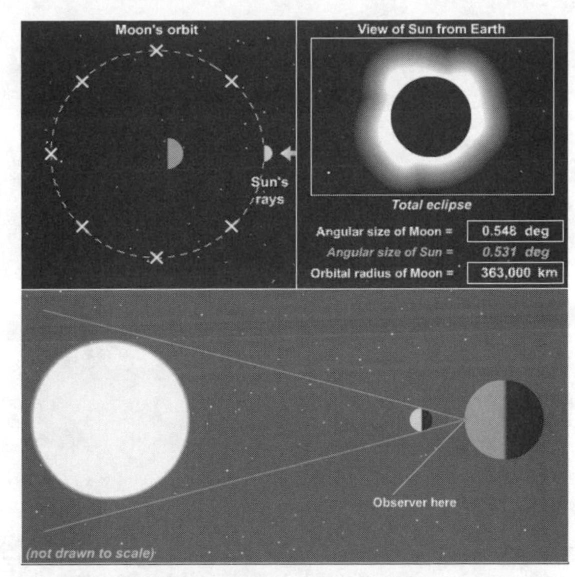

Figure 3-4 This tool from Lesson 2 allows you to observe what a solar eclipse looks like depending on the Moon's position in its orbit.

Position Number	Angular Size of Moon (in degrees)	Orbital Radius of Moon (in km)	Total or Annular Eclipse?
1			
2			
3			
4			
5			
6			
7			
8			

In general, how does the angular size of the Moon appear to vary with changes in its orbital radius?

According to the table, ANNULAR | TOTAL *(circle one)* solar eclipses seem more likely.

In which position would the eclipse be least noticeable to observers on the Earth? Explain your reasoning.

MASTERINGASTRONOMY WEB TUTORIALS

4

Phases of the Moon

Your goals in this tutorial are to:

- understand the cause of lunar phases
- predict how the Moon looks from Earth in a particular phase and when it will rise and set
- determine the orbital position of the Moon based on where it is in the sky at a given time

(In this tutorial it will be especially helpful to use the tools in each lesson to help you answer the questions. You will learn much more this way than by trying to answer the questions from memory.)

LESSON 1

1. When the Moon is between the Earth and the Sun, the Moon's DARK | ILLUMINATED *(circle one)* side is facing the Earth, and the Moon will be visible from the Earth mostly during the DAY | NIGHT *(circle one)*.

2. When the Earth is between the Moon and the Sun, the Moon's DARK | ILLUMINATED *(circle one)* side is facing the Earth, and the Moon will be visible from the Earth mostly during the DAY | NIGHT *(circle one)*.

Figure 4-1 This tool for the Lesson 1 questions shows the Moon orbiting the Earth and being illuminated by the Sun.

3. The same side of the Moon is always facing the Earth. This side of the Moon is ALWAYS DARK | ALWAYS ILLUMINATED | COULD BE EITHER DARK OR ILLUMINATED OR PARTLY BOTH *(circle one)*.

4. Use the tool to determine if the Moon can ever be seen from Earth during the day.

LESSON 2

5. What time is it when your location on the Earth is facing directly toward the Sun?

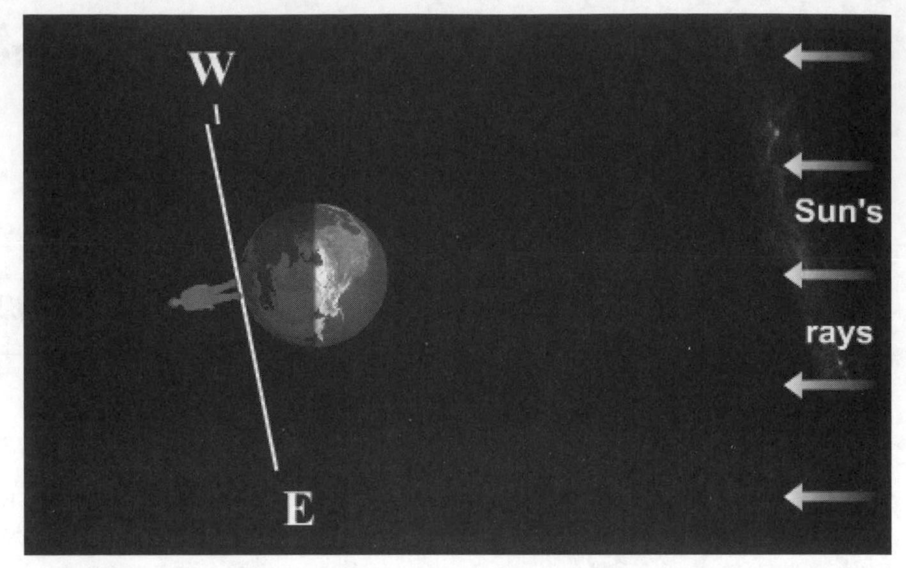

Figure 4-2 This tool for the Lesson 2 questions shows how an observer's location on Earth defines the time of day.

6. What time is it when your location on the Earth is facing directly opposite the Sun?

7. As seen from above the North Pole, the Earth rotates CLOCKWISE | COUNTERCLOCKWISE *(circle one)*. This is from EAST TO WEST | WEST TO EAST *(circle one)*.

LESSON 3

8. When does the new Moon rise and set?

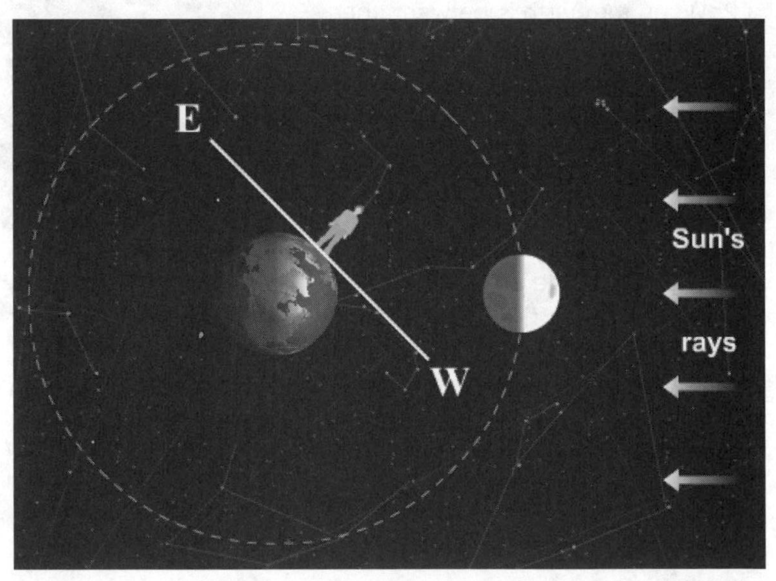

Figure 4-3 Use this tool for the Lesson 3 questions to determine when the Moon is visible from Earth.

9. If the Moon is up all night, what phase is it in?

10. If the Moon is up half the night, then half the day, what phase is it in?

11. When does a first-quarter Moon rise and set?

12. If the Moon sets just after the Sun, what phase is the Moon in?

LAB ACTIVITY Phases of the Moon

The goal in this activity is to determine the phase and the rising and setting times of the Moon when observed from the Earth.

Use the tool shown in Figure 4-4, taken from the exercises section of the tutorial, to answer the questions below.

About what time is it in Figure 4-4?

Where will the Moon be in the observer's sky in Figure 4-4?

RISING | HIGH IN THE SOUTH | SETTING *(circle one)*

In what phase is the Moon in Figure 4-4?

At what times will the Moon rise and set in Figure 4-4?

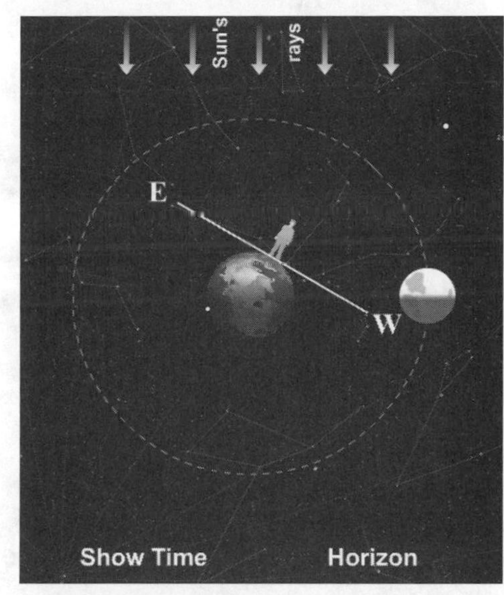

Figure 4-4 Shown here is the lunar phases tool from the Exercises section.

What is the name of the phase shown in Figure 4-5?

At what times will the Moon rise, set, and be highest overhead when in this phase?

If the Moon is at its highest in the south at 9 A.M., what phase is it in?

Figure 4-5 This photo shows a particular phase of the Moon.

For the phase in the previous question, draw where the Moon would be in its orbit in Figure 4-4.

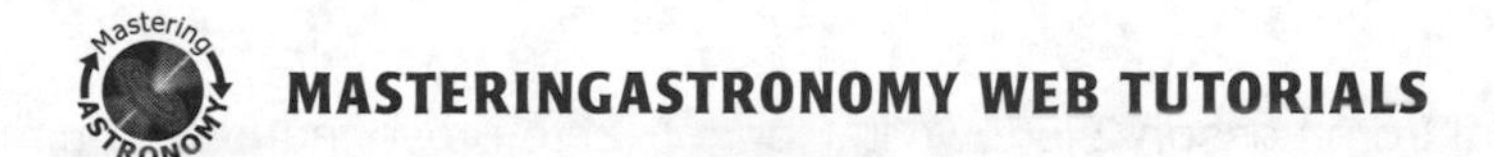

5
Orbits and Kepler's Laws

Your goals in this tutorial are to:

- understand that objects in orbit are simply falling
- to be able to explain Kepler's laws of planetary motion

LESSON 1

1. What is necessary for an object to be able to escape a planet's orbit?

2. What effect does an object's mass have on its orbit (see Figure 5-1)? Try different masses while keeping the speed constant.

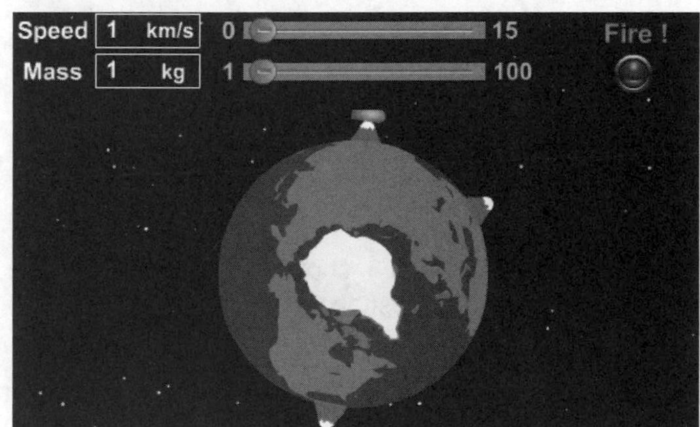

Figure 5-1 This tool in Lesson 1 determines the effect of launch speed and mass on an object's orbit.

3. Explain why a feather and a hammer fall at the same rate on the Moon.

4. Why do they fall at different rates on the Earth?

LESSON 2

5. Use the ellipse tool from Lesson 2 (shown in Figure 5-2) to prove that the semimajor axis of an ellipse is a planet's average orbital radius:

   A. Orbital radius when the planet is closest to the Sun: perihelion = ________________ AU

   B. Orbital radius when the planet is farthest from the Sun: aphelion = ________________ AU

   C. Average of perihelion and aphelion = ________________ AU

   D. Length of the major axis of the ellipse = ________________ AU

   E. Length of the semimajor axis of the ellipse = ________________ AU

   Are the average orbital radius (C) and the semimajor axis (E) equal?

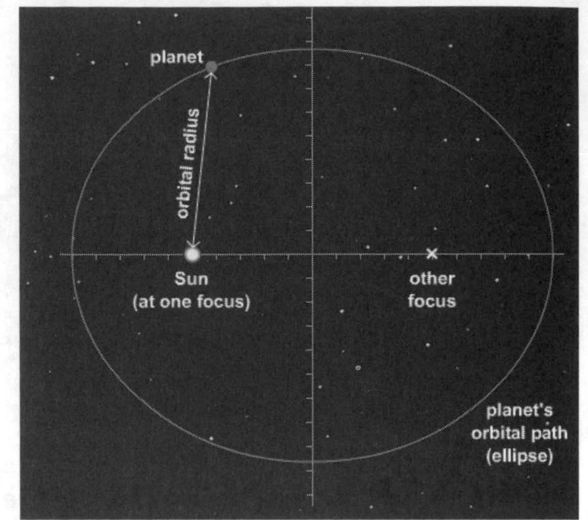

Figure 5-2 Shown here is the ellipse tool from Lesson 2.

6. What special case of an ellipse is a circle?

LESSON 3

Use the tool from Lesson 3, shown in Figure 5-3, to answer the following questions.

7. When a comet is closer to the Sun, its orbital speed is FASTER | SLOWER *(circle one)*. This means that orbital radius and speed are DIRECTLY | INVERSELY *(circle one)* proportional.

8. Explain why we see Comet Halley from the Earth for a few months only once every 76 years.

9. Is orbital radius a more meaningful quantity for the Earth or for Comet Halley? Explain your answer.

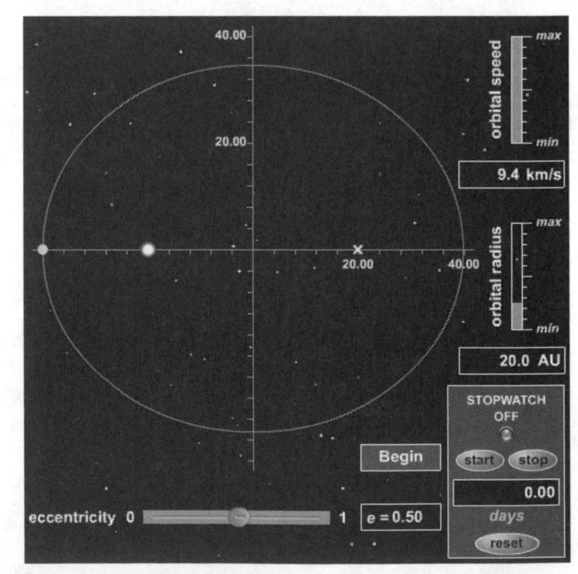

Figure 5-3 Use this tool from Lesson 3 to observe an orbit after setting its eccentricity.

10. The orbital period is the amount of time a planet takes to go around the Sun. Which of the following quantities would the orbital period of a planet depend on? *(circle your answers)*

    Orbital Speed   Orbital Radius   Eccentricity

LESSON 4

Use the tool from Lesson 4, shown in Figure 5-4, for questions 11 and 12.

11. Use the tool to predict the orbital period of an asteroid that has an orbital radius of 4 AU.

    $P =$ ______

    Now use Kepler's third law, $P^2 = a^3$, to check your prediction.

    $a^3 =$ ______________. Now take the square root of the result, $P =$ ______________.

    Is the orbital period the same as the answer you calculated?

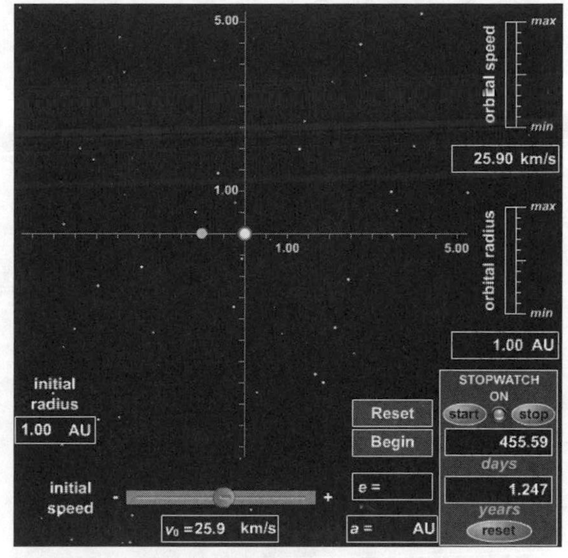

Figure 5-4 This tool from Lesson 4 determines an object's orbital period after the initial orbital radius is set.

12. Use the tool to predict the orbital radius of an asteroid that has a period of two years. *Hint:* Would it be farther or closer to the Sun than Earth? Check your answer using Kepler's third law.

LAB ACTIVITY Orbits and Kepler's Laws

The goal of this activity is to test Kepler's third law.

For each semimajor axis given in the table, use the stopwatch in the tool from Lesson 4 (shown below in Figure 5-5) to time the orbital period. Then test Kepler's third law.

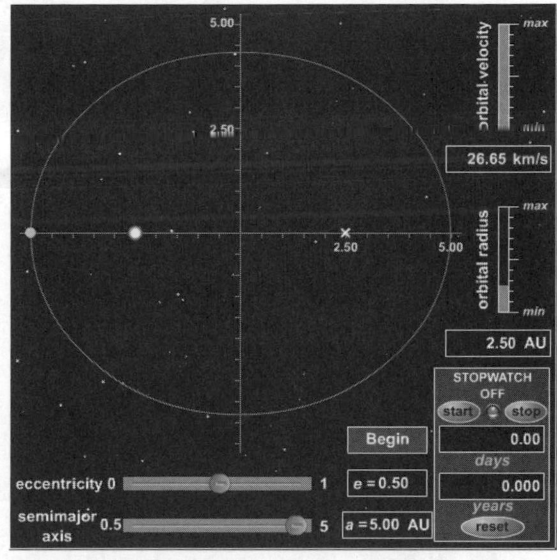

Figure 5-5 This tool from Lesson 4 determines an object's orbital period after the orbit's eccentricity and semimajor axis are set.

Semimajor Axis (in AU)	Period (in years)	P^2 (period squared)	a^3 (semi-major axis cubed)
1			
2			
3			
4			
5			

Do the squares of the periods seem to come out close to equal to the cubes of the semimajor axes?

Why do you think some of the squared periods and cubed semimajor axes came out very close but not exactly the same?

MASTERINGASTRONOMY WEB TUTORIALS

6

Motion and Gravity

Your goals in this tutorial are to:

- discuss how force, mass, and acceleration are related using Newton's laws of motion
- describe how your weight depends on your mass, Earth's mass, and the radius of Earth
- discuss which factors determine the force of gravity between a planet and a star

LESSON 1

You should use the tools in Lesson 1 to help you answer questions 1–3.

1. If the amount of force on a mass increases, how does the acceleration change? Explain your answer.

2. How do the accelerations compare if the same force is exerted first on a large mass and then on a small mass? Explain your answer.

3. If you want the same acceleration, how must you change the force exerted on a small mass compared to a large mass? Explain your answer.

4. If two objects of different mass are dropped from the same height at the same time, ignoring air resistance, which will hit the ground first, the more or the less massive object? (Use the tool from Lesson 1 shown in Figure 6-1 for help.) Explain your answer.

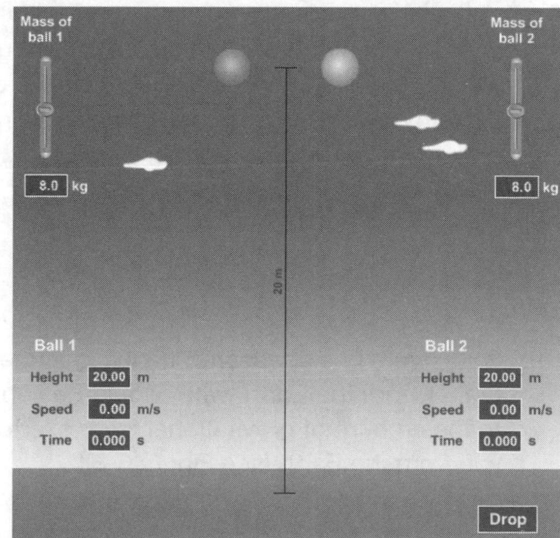

Figure 6-1 This tool from Lesson 1 simulates free falling masses.

5. If you didn't ignore air resistance in question 4, how do you think your answer would be affected? Explain your reasoning.

LESSON 2

6. Explain the difference between mass and weight.

You may wish to use the tool from Lesson 2, shown in Figure 6-2, to help you answer questions 7–9.

7. If you are transported to a planet that has the same mass as Earth, but is larger in size, how would this affect your weight when standing on the planet's surface? Explain your answer.

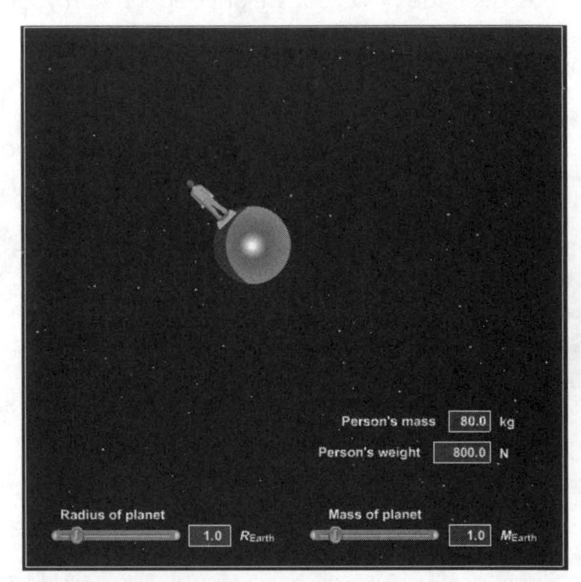

Figure 6-2 This tool from Lesson 2 determines the weight of a mass on planets of different radii and mass.

8. If you are transported to a planet that is the same size as Earth, but has less mass, how would this affect your weight when standing on the planet's surface? Explain your answer.

9. If you could stand on the surface of Uranus, your weight would be very similar to what it is on Earth. Explain how this is possible, given that Uranus is both more massive and larger than Earth.

10. If you were in a spacecraft traveling farther and farther from the Earth, what would be happening to the amount of gravitational force exerted on you from the Earth (see Figure 6-3)?

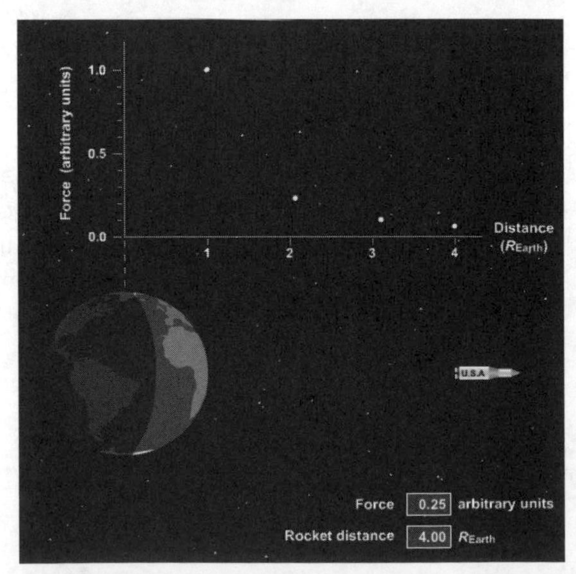

Figure 6-3 The gravitational force exerted on a rocket changes as the rocket moves away from Earth.

11. If you are in a spacecraft midway between the Earth and the Moon, which object would have a stronger gravitational pull on your ship? Explain your answer.

12. Would it ever be possible for the Moon to exert more gravitational pull on your ship than the Earth would? Explain your answer.

LAB ACTIVITY Motion and Gravity

The purpose of this activity is to learn how the radius and mass of a planet affect the gravitational pull it exerts on another mass and to demonstrate the difference between mass and weight.

Use the tool from Lesson 2 (shown in Figure 6-4) to determine both the person's largest and smallest possible weights. Record your results in the table below.

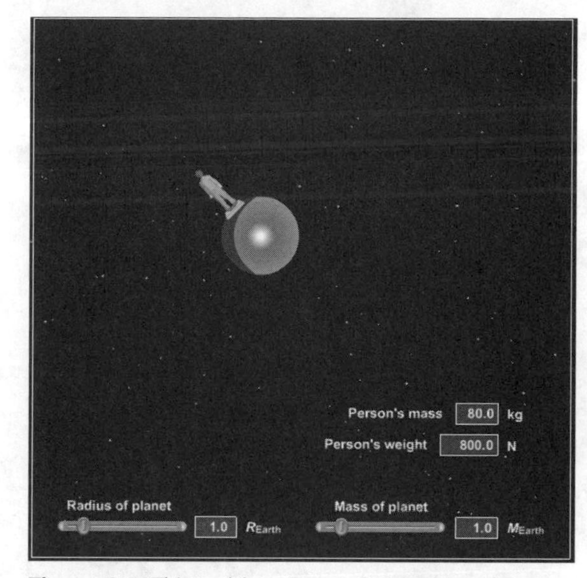

Figure 6-4 This tool from Lesson 2 determines the weight of a mass on planets of different radii and mass.

Person's Mass (kg)	Radius of Planet (R_{Earth})	Mass of Planet (M_{Earth})	Person's Weight (N)
80			Largest:
80			Smallest:

Which extremes of the radius and mass of the planet were needed to get the largest possible weight for the person?

Which extremes of the radius and mass of the planet were needed to get the smallest possible weight for the person?

7

Energy

Your goals in this tutorial are to be able to:

- list the different forms of energy, giving an example of each type
- explain how kinetic energy depends on speed and mass
- explain how a spaceship's gravitational potential energy depends on its distance from a planet, the mass of the planet, and the ship's mass
- explain the meaning of conservation of energy

LESSON 1

1. On what property or properties of an object does kinetic energy depend?

2. On what property or properties of an object does gravitational potential energy depend?

3. Describe the changes that occur in an object's potential energy when it is thrown upward, reaches a maximum height, and then returns to the position from which it was thrown.

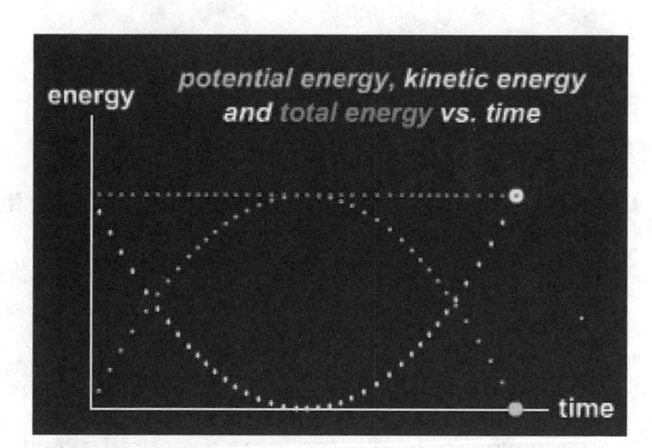

Figure 7-1 Changes in potential, kinetic, and total energy for an object that is thrown upward, reaches a maximum height, and then returns to its initial position. Use this figure to help answer questions 3–5.

4. Describe the changes that occur in an object's kinetic energy when it is thrown upward, reaches a maximum height, and then returns to the position from which it was thrown.

5. Describe the changes that occur in an object's total energy when it is thrown upward, reaches a maximum height, and then returns to the position from which it was thrown.

6. What condition relating to energy must be met for an object launched from Earth to escape Earth's gravity?

LESSON 2

7. What property of an object is a measurement of its thermal energy?

8. How will the radiation from a hotter object differ from the radiation of a cooler object?

9. On what does electric force depend that is analogous to mass in gravitational force?

10. How will two objects with gravitational potential energy between them tend to move?

11. How will two objects with electrical potential energy between them tend to move (see Figure 7-2)?

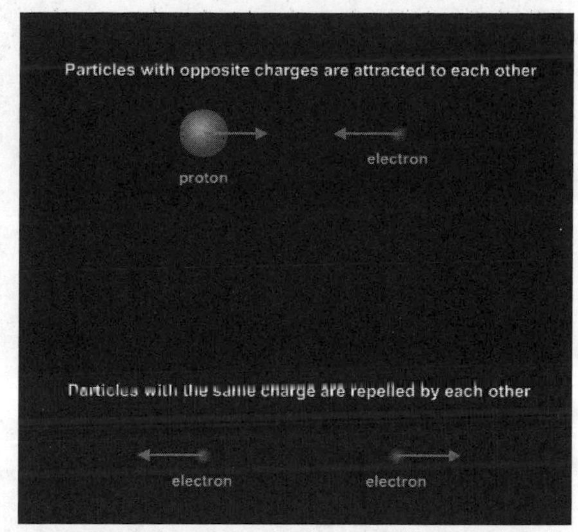

Figure 7-2 Charged particles may attract or repel one another.

12. Write out Einstein's mass-energy equivalence equation. Explain the meaning of each symbol in the equation and how the equation shows that nuclear reactions are a source of tremendous amounts of energy.

LAB ACTIVITY Energy

Use the tool from Lesson 1 (shown in Figure 7-3) to complete the table and answer the questions below.

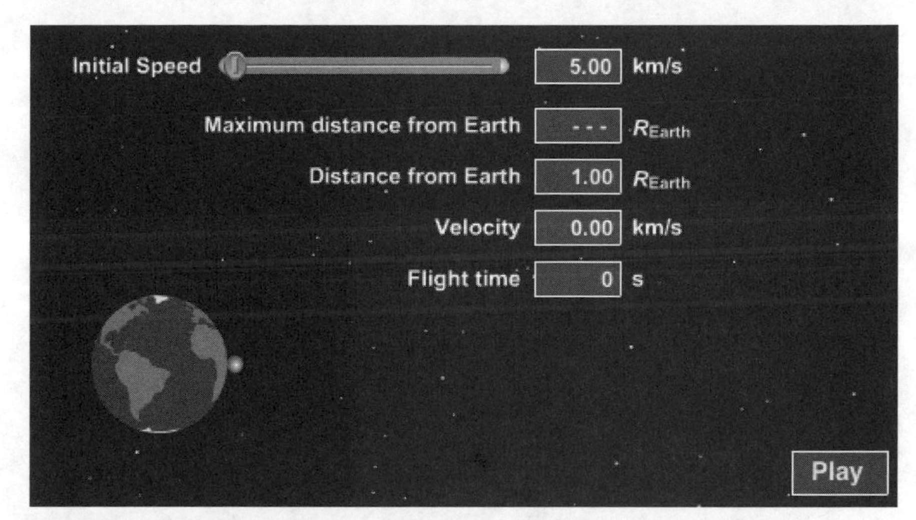

Figure 7-3 This tool from Lesson 1 determines the distance traveled and flight time for a rocket launched from Earth at different speeds.

Initial Speed (km/s)	Maximum Distance from Earth (R_e)	Flight Time (s)
5		
6		
7		
8		
9		
10		
11		
12		
13		
14		
15		

What is the minimum speed necessary to escape Earth's gravity?

If escape speed is achieved, what will happen to the values of the maximum distance from Earth and the flight time?

Explain in terms of kinetic and potential energy what happens when escape speed is achieved.

8
Light and Spectroscopy

Your goals in this tutorial are to:

- define wavelength and list the types of electromagnetic waves in order of decreasing wavelength
- discuss how spectra are used in astronomy
- explain how emission and absorption lines are produced and discuss the information they convey
- describe how to use the spectrum emitted by an object to determine the object's temperature

LESSON 1

1. Which property of a light wave determines the color we see?

2. What do we see that the amplitude of a wave produces?

3. Name a type of electromagnetic wave that has a long wavelength. Name one that has a high frequency (see Figure 8-1).

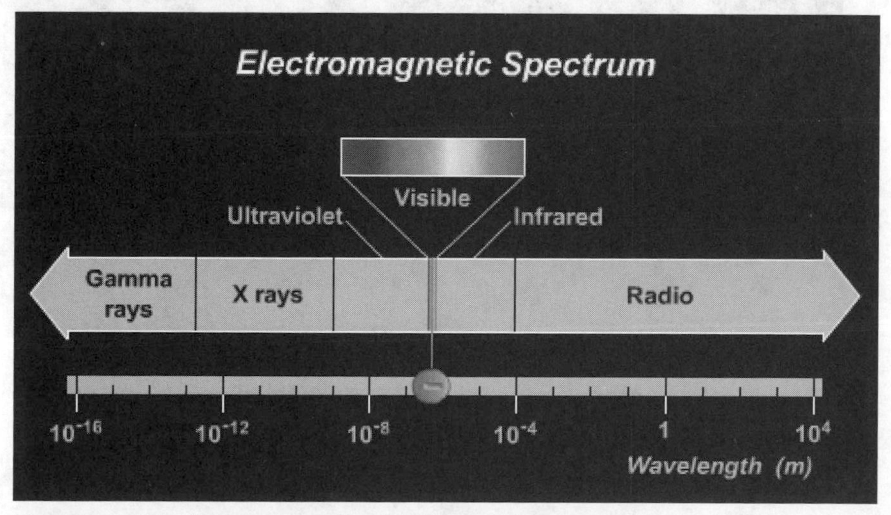

Figure 8-1 This diagram shows the electromagnetic spectrum.

LESSON 2

4. As you learned in this lesson, some objects give off or reflect more than one color of light. What determines the color that we see when we look at an object?

5. Of the five objects in the tool shown in Figure 8-2, which does not give off its own light?

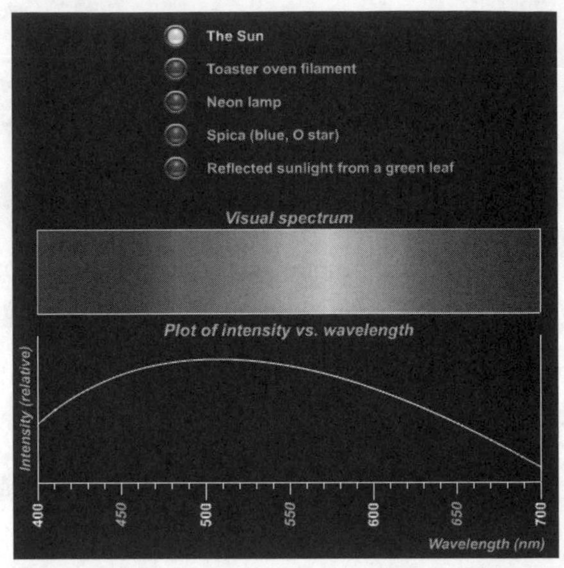

Figure 8-2 This tool from Lesson 2 displays spectra of different objects.

6. Does a blue star, such as Spica, give off any other colors? Why does it appear to be blue?

LESSON 3

7. Which colors do you see in the hydrogen spectrum?

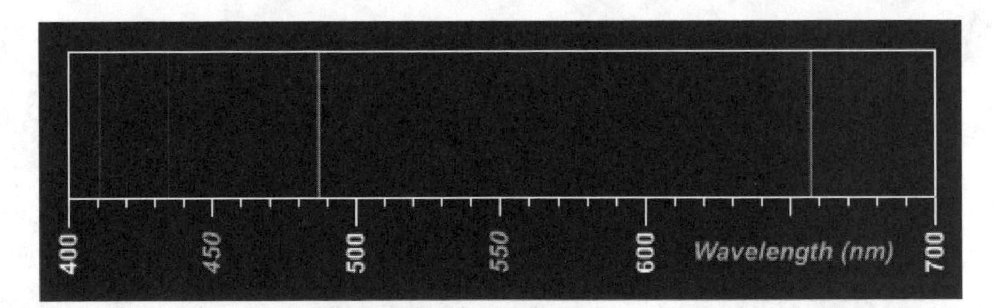

Figure 8-3 This diagram shows the spectrum of hydrogen.

8. Which color is produced by the emission of the highest energy photons? The lowest energy photons?

9. Which element's atomic spectrum was the simplest of the ones you looked at? What does this suggest about the structure of that kind of atom?

LESSON 4

10. What kind of spectrum does a solid object emit (see Figure 8-4)?

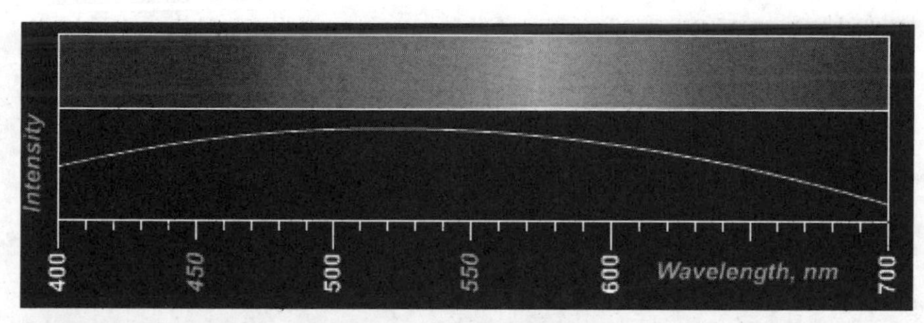

Figure 8-4 The spectrum emitted from a solid object appears as shown.

11. Which kinds of electromagnetic waves do hotter objects emit? Cooler ones?

12. How would the temperature of a blue star compare to that of a red star?

LAB ACTIVITY Light and Spectroscopy

The goal of this activity is to show how the temperature of an object and the wavelength of the light it emits are related.

Use the tool from Lesson 4 (Figure 8-5) to answer the questions.

What color star would have the spectrum shown in Figure 8-5?

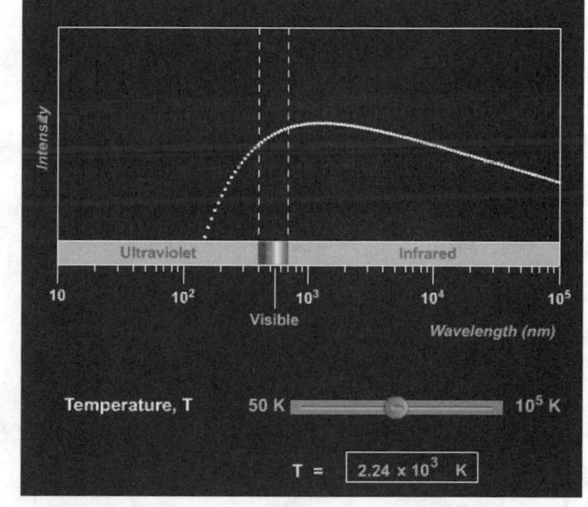

Figure 8-5 The graph shows the intensity of light of different wavelengths given off by objects of different temperatures.

Adjust the temperature until the spectrum gives off all the visible colors with about the same intensity. At about what temperature does this occur?

What color would a star with this spectrum be?

What is the maximum temperature of an object that gives off no visible light?

In what part of the spectrum does a star that looks blue probably have its peak wavelength?

9
The Doppler Effect

Your goals in this tutorial are to:

- understand the cause of the Doppler effect
- determine whether light will be redshifted or blueshifted
- discuss how the Doppler effect is used to determine the distances to astronomical objects
- discover the limitations in using the Doppler effect to determine an object's speed

LESSON 1

1. As the source of a sound passes you, which two properties will you hear change?

2. Of the properties in question 1, which property changes are due to the Doppler effect?

3. Which property of a sound wave is responsible for the pitch that you hear?

4. If a race car is approaching you, what change will you hear in the pitch of the motor (see Figure 9-1)?

5. Explain why this change in pitch occurs.

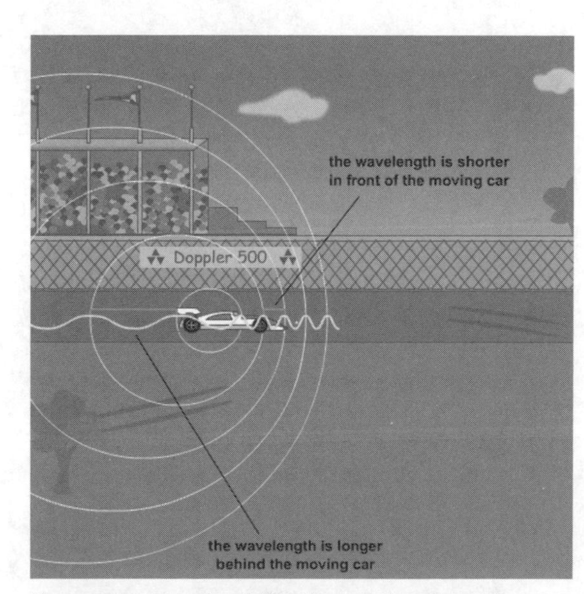

Figure 9-1 Doppler shift in the wavelengths of the sound varies in front of and behind a moving car.

6. Which aspect of the light that you see is caused by the same wave property that causes the pitch of the sound that you hear?

7. Which wave property causes the color of light you see?

LESSON 2

8. Explain what is meant by a redshift and a blueshift in a spectrum. Explain why the shifts occur and what they tell us about their light sources.

9. If two galaxies are redshifted, one more than the other, what does that tell you about the direction in which they are moving? Their speeds?

10. If something is blueshifted, does that necessarily mean that it appears blue? Explain your answer.

11. If a star shows no Doppler shift, does that necessarily mean that it is not in motion relative to us? Explain your answer.

12. A police officer pulls you over for running a red light. You claim that your motion caused a Doppler shift and you thought the light was green! She still gives you a ticket, but not for running a red light. What is the ticket for? Explain your answer. (Figure 9-2 may give you a hint.)

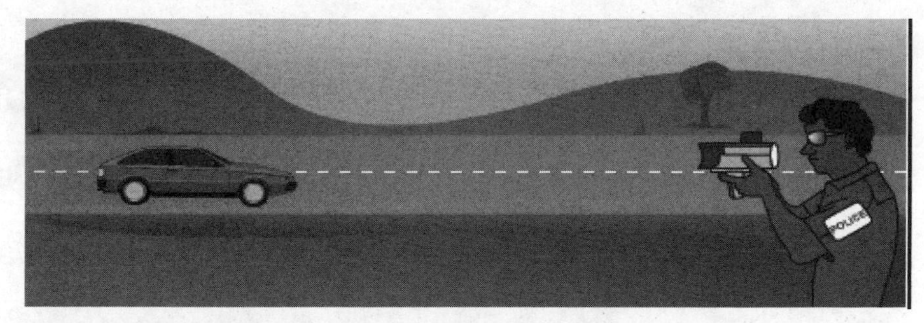

Figure 9-2 What is the police officer using the Doppler effect to measure?

LAB ACTIVITY The Doppler Effect

The goal of this activity is to observe changes that occur in light due to the Doppler effect.

Use the tool from the end of Lesson 1 (shown in Figure 9-3) to complete the table. Try to adjust the speed of the helicopter so that you observe a wavelength in each portion of the visible spectrum. Record the speeds, using a + for a speed moving toward you and a − for a speed moving away from you.

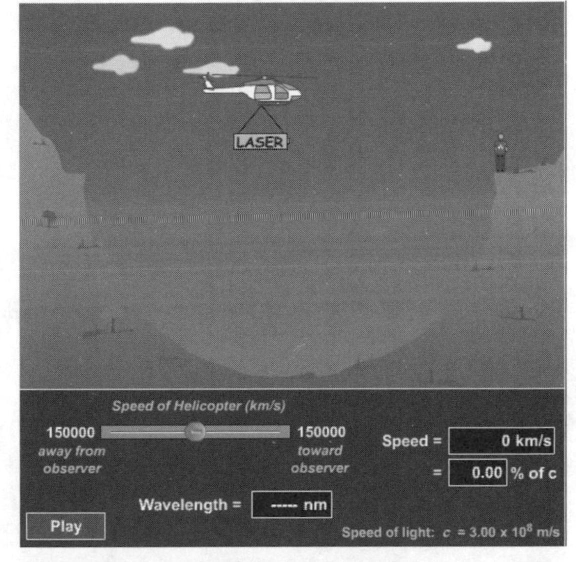

Figure 9-3 Doppler shifts in the LASER light from the helicopter depend on its speed and direction.

Color	Wavelength (nm)	Speed (km/s)
Violet	400–450	
Blue	450–500	
Green	500–550	
Yellow	589	0
Orange	600–650	
Red	650–700	

What color is the laser when the helicopter is standing still?

Name a color that the laser could be when the helicopter is moving away from you.

Name a color that the laser could be when the helicopter is moving toward you.

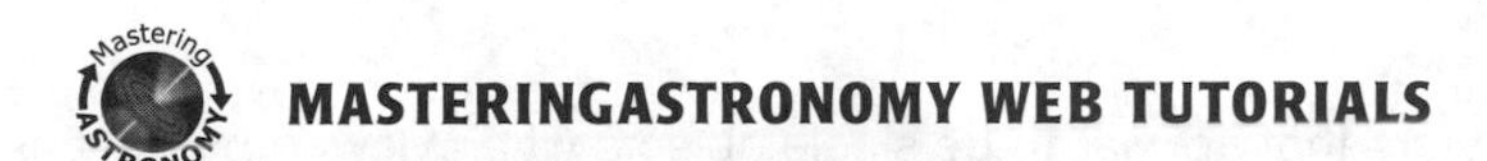

10 Telescopes

Your goals in this tutorial are to:

- explain how reflecting telescopes focus light and produce images
- show how light-gathering power and angular resolution depend on the diameter of the telescope's primary mirror
- explain the main advantage of putting a telescope above the Earth's atmosphere

LESSON 1

1. If stars radiate their light in all directions, explain why their rays are essentially parallel when they reach telescopes on Earth.

2. Why do astronomers use a secondary mirror in a reflecting telescope?

3. If starting from the left you see with your eyes a blue and then a yellow star, in what order starting from the left will the stars appear when viewed through a telescope? Explain your reasoning (see Figure 10-1).

4. If you double the diameter of the mirror of a reflecting telescope, how does the amount of collected light change?

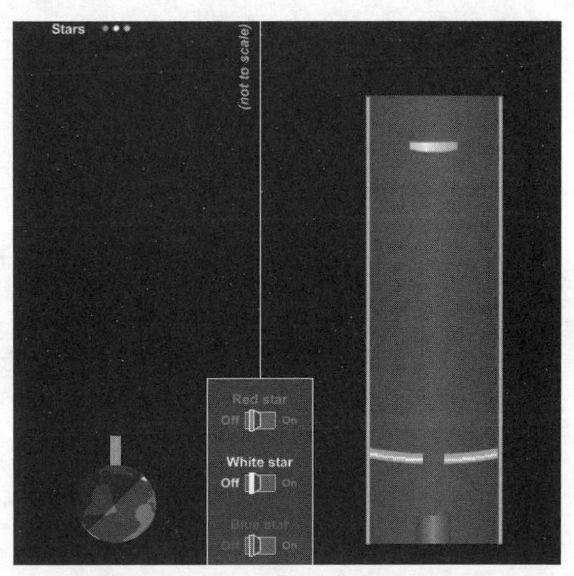

Figure 10-1 This tool in Lesson 1 shows where an image will appear on a detector.

LESSON 2

5. Why are so many stars that appear to be single objects when viewed by the naked eye actually seen to be pairs of stars when viewed through a telescope?

6. How does diffraction through a larger hole change the size of an image?

7. How does the angular resolution of a telescope change with the size of the mirror? Does a larger or a smaller mirror have better angular resolution? Explain your reasoning.

8. How does the angular resolution of a telescope compare when observing a red star to when observing a blue star? For which color star is the angular resolution better? When it comes to angular resolution is bigger, better? Explain your answer.

9. Using the tool at the end of Lesson 2 (see Figure 10-2), determine which will have greater angular resolution: a 0.1 m diameter ultraviolet telescope observing light of wavelength 2×10^{-7}m or a 10 m infrared telescope observing light of wavelength 1×10^{-5} m.

   Which telescope will have greater resolution? By how many times? In this case, which telescope has better resolution? Explain your reasoning.

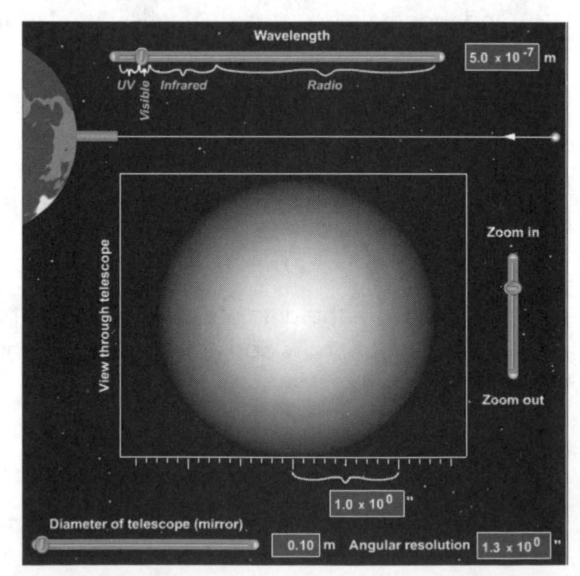

Figure 10-2 This tool illustrates the angular resolution of telescopes of varying diameters at different wavelengths of light.

LESSON 3

10. Which types of electromagnetic waves penetrate the Earth's atmosphere the best (see Figure 10-3)?

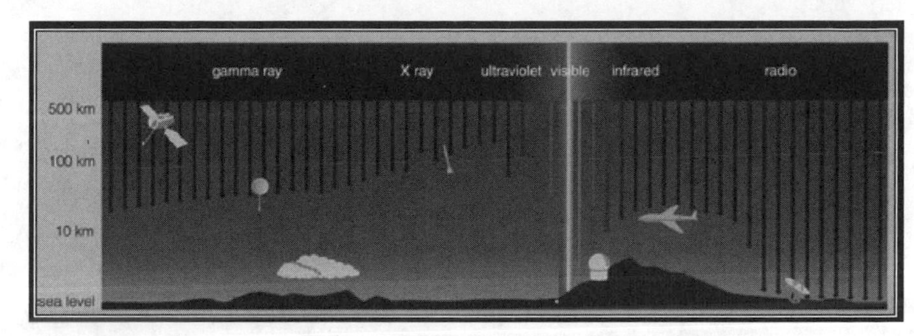

Figure 10-3 Shown are the electromagnetic waves in the Earth's atmosphere.

11. Explain the advantage(s) of placing a telescope on a mountain.

12. Explain the advantage(s) of placing a telescope in space.

LAB ACTIVITY Telescopes

The purpose of this activity is to determine how the amount of light collected by a telescope mirror varies with the size of the mirror.

Use the tool from Lesson 1 (Figure 10-4) to complete the table below and plot the graphs.

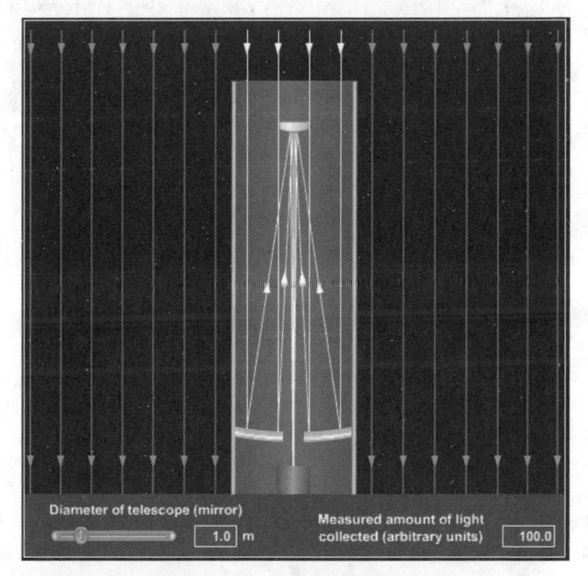

Figure 10-4 This tool from Lesson 1 illustrates the amount of light collected by telescopes of different sized mirrors.

Diameter of Mirror (m)	Square of Diameter	Amount of Light Collected
1		
2		
3		
4		
5		

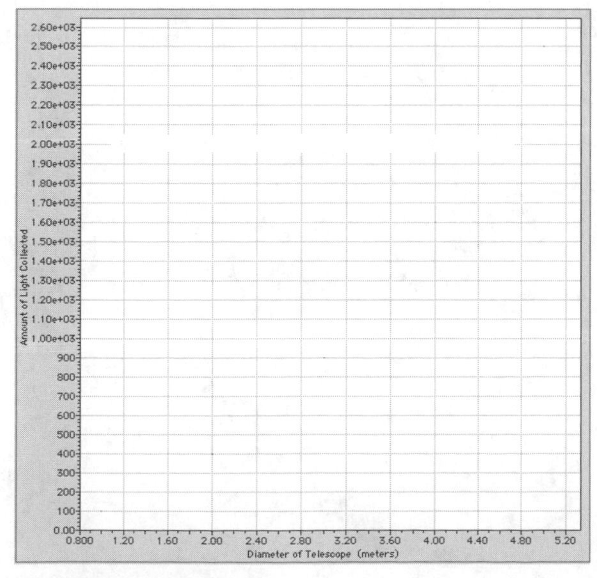

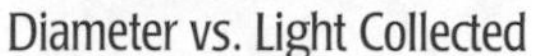

Diameter vs. Light Collected

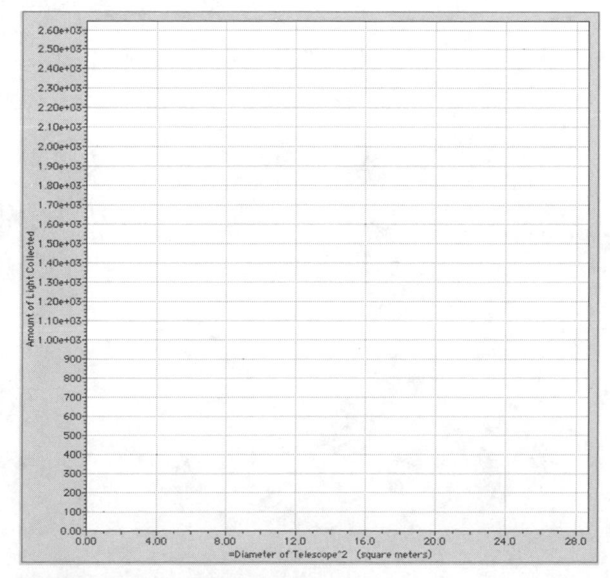

Diameter Squared vs. Light Collected

The amount of light collected by a telescope mirror is directly proportional to the *(circle one)* DIAMETER | SQUARE OF THE DIAMETER.

Which best describes the relationship between the amount of light collected (L) and the diameter of a telescope mirror (d)? *(circle one) Hint*—were either of your graphs a straight line?

$L = d$ $\quad$ $L = d^2$ $\quad$ $L = 1/d$ $\quad$ $L = 1/d^2$

11 Formation of the Solar System

Your goals in this tutorial are to:

- list the sequence of events in the formation of the solar system
- list the similarities and differences between the types of planets and give the reasons for them
- discuss the reasons for the orbital properties of the solar system

LESSON 1

See Figure 11-1 for help with the questions in this lesson.

1. Name the two major types of planets in our solar system.

2. Name the members of each of the two types of planets.

3. List several differences between the two types of planets.

Jovian Planets	Terrestrial Planets
• Larger size and mass • Low density (composed mostly of hydrogen and helium gases and, in some cases, hydrogen compounds) • Farther from the Sun • Cooler	• Smaller size and mass • High density (composed mainly of rock and metal) • Closer to the Sun • Warmer

Figure 11-1 This illustration shows the two planetary types and their properties.

4. How do the orbital distances from the Sun compare for the two types of planets?

LESSON 2

5. Explain why a molecular cloud flattens out as it collapses.

6. How does the temperature of the cloud vary with the distance from its center?

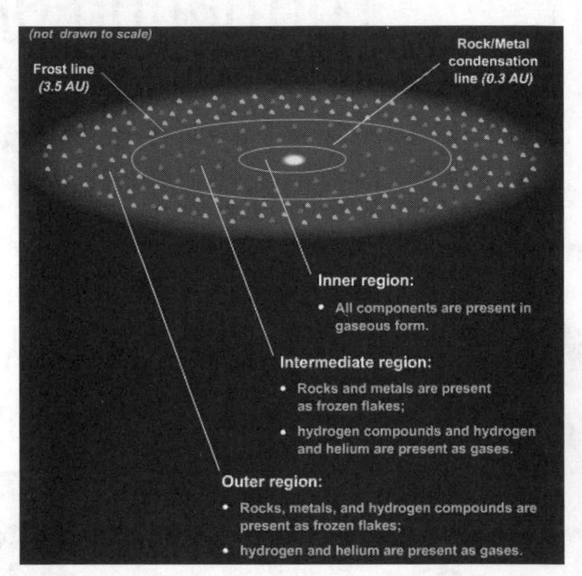

Figure 11-2 This diagram illustrates materials found in the solar nebula and their locations. This tool will help you answer questions 7 and 8.

7. Explain why the frost line is the boundary between the two types of planets.

8. Why are no planets found within 0.3 AU of the Sun?

LESSON 3

9. Explain why the planetesimals beyond the frost line were initially able to grow larger than those inside the frost line.

10. Which of the two planet types has evolved more from its initial condition?

11. Why were planets closer to the Sun not able to retain lighter gases (see Figure 11-3)?

12. What do we now call the left-over planetesimals (those that did not become part of a planet)?

Figure 11-3 Planets closer to the Sun could not retain lighter gases, while those farther from the Sun could.

LAB ACTIVITY Formation of the Solar System

The purpose of this activity is to see the effect on planetary formation of temperatures at various locations in the solar nebula.

Mark the positions of the rock/metal condensation line and the frost line on Figure 11-4. Also on Figure 11-4, indicate in which region each of the two major types of planets formed. You can find this tool in Lesson 2.

What kinds of planets (if any) formed inside the rock/metal condensation line? Explain the reason for the kind of planet that formed there, or, if there are no planets, the reason they did not form.

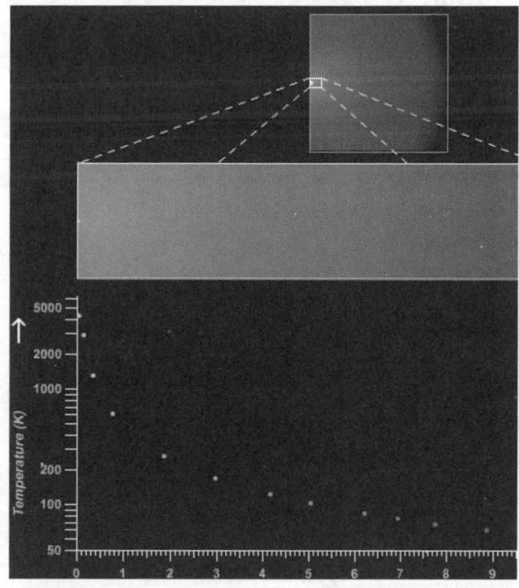

Figure 11-4 Shown here is a plot of the temperature of the solar nebula as a function of distance from the Sun.

What kinds of planets (if any) formed between the rock/metal condensation line and the frost line? Explain the reason for the kind of planet that formed there, or, if there are no planets, the reason they did not form.

What kinds of planets (if any) formed beyond the frost line? Explain the reason for the kind of planet that formed there, or, if there are no planets, the reason they did not form.

Which major type of planet is most evolved (i.e., which type has gone through the most changes)? Explain how this can be attributed to where these planets formed.

12
Shaping Planetary Surfaces

Your goals in this tutorial are to:

- list the four main geological processes that can shape a planet's surface
- identify features caused by these processes
- explain what factors determine which geological processes will be important in shaping a planet's surface
- predict the final appearance of a planetary surface based on the size and temperature of the planet

LESSON 1

1. What are the four geologic processes that are responsible for shaping planetary surfaces (see Figure 12-1)?

Figure 12-1 This illustration shows the geologic processes that shape a planetary surface.

2. What factor controls the size that an impact crater will have?

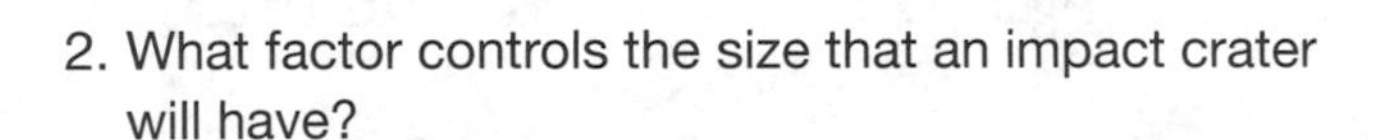

3. What major component of a planet are gases emitted from volcanoes responsible for forming?

LESSON 2

4. How has the rate of crater formation in the solar system changed over time? Why has it changed?

5. Why are there so many more impact craters on the Moon and Mercury than on the Earth, *even though Earth is a bigger target*? Why are the craters on the Moon and Mercury so well preserved?

6. Which geologic processes are controlled by a planet's internal temperature? What factor affects how long these processes will continue on a planet (see Figure 12-2)?

7. Why, unlike most other planets, is erosion a dominant process on Earth?

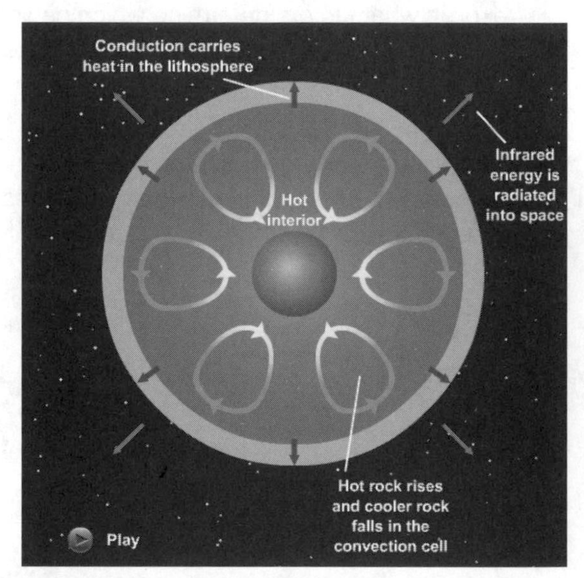

Figure 12-2 Use this animation for Lesson 2 to help you answer question 6.

LESSON 3

8. Which geologic process occurs at the same rate on all planets?

9. What factors control the geologic evolution of a planet?

10. Which factor seems to control the types of geologic activity that occur on a planet?

11. Which property of a planet seems to control the amount of time that the geologic process continues on a planet?

12. What geologic process(es) would dominate on a small, cold planet?

LAB ACTIVITY Shaping Planetary Surfaces

The goals of this activity are to learn which geologic process will dominate the evolution of planetary surfaces of various temperatures and sizes and how long the processes will last.

For this activity, use the tool from Lesson 3 shown in Figure 12-3.

By changing the planet temperature and planet size, you can simulate the evolution of a planetary surface over time. In the table below, enter the relative temperature (hot, medium, or cold) and size (large, medium, or small) that you think represents the specified objects in our solar system.

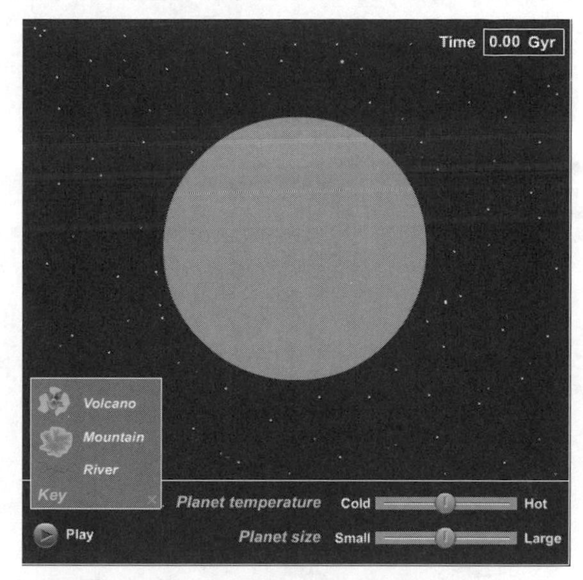

Figure 12-3 This tool from Lesson 3 shows the evolution of a planetary surface after the planet's temperature and size are selected.

Object	Temperature	Size
Mercury		
Venus		
Earth	Medium	Large
Moon		
Mars		
Asteroid*		

*Note that the largest asteroids are only about 500 miles across and that most of them are located beyond the orbit of Mars.

Once you have decided on the temperature and size that will be used for each planet, use the tool to set the planet temperature and planet size and click on play to observe the evolution of each planet's surface. For each planet, notice which processes occur, when they occur, and how long changes in the planet's surface persist. After you observe each simulation, enter what you observed in the table below. Then, based on your observations, answer the questions that follow.

Object	Time*	Dominant Process(es)	Realistic?
Mercury			
Venus			
Earth	4.6 G-years	Tectonic, volcanic, erosion	yes
Moon			
Mars			
Asteroid			

*Enter the amount of time the evolution took to stop or slow down considerably; be patient enough to allow the tool to run until it stops on its own.

If necessary, rerun some of the simulations with the tool to answer the questions.

What did the objects dominated by impact cratering seem to have in common? Did other factors seem to affect their surface evolution?

Which objects seemed to take the longest to evolve? Which evolved fastest? Which factor seems to have the greatest effect on this?

What feature did you observe on the Earth that you did not find on Venus? What is the reason for this? Did you observe this feature on Mars?

Which factor seemed to be the main reason that Mars evolved differently than the Earth did?

Which process seemed to dominate early on, but tapered off during the evolution of the planetary surfaces?

MASTERINGASTRONOMY WEB TUTORIALS

13

Surface Temperature of Terrestrial Planets

Your goals in this tutorial are to be able to:

- explain how distance from the Sun, reflectivity, and atmosphere affect the surface temperature of a terrestrial planet
- give the main reason that Venus is so much hotter and Mars is so much colder than the Earth.

LESSON 1

Use the tools in this lesson to answer the following questions.

1. On which factors does the rate at which a terrestrial planet emits radiation depend?

2. Does the size of a planet affect its temperature? Why or why not?

3. How is the change in temperature between day and night affected by the rotation rate of a terrestrial planet? Explain your reasoning (see Figure 13-1).

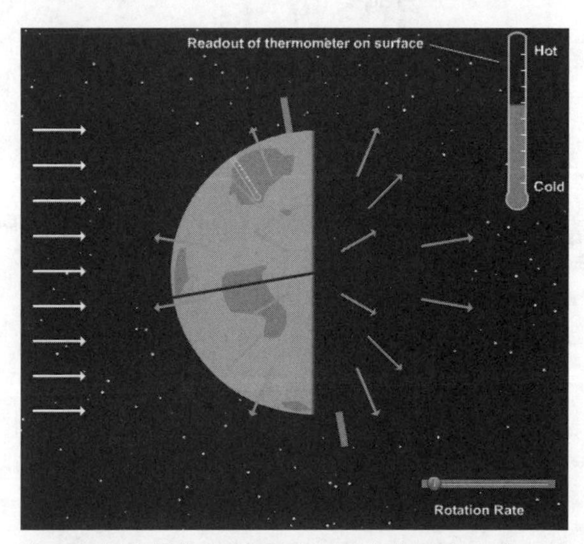

Figure 13-1 This tool from Lesson 1 shows how the rotation rate of a planet affects its temperature.

LESSON 2

4. Use the tool in this lesson (see Figure 13-2) to predict the Earth's surface temperature. The Earth is 1 AU from the Sun. Also use the tool to predict Mercury's surface temperature. Mercury is about 0.4 AU from the Sun.

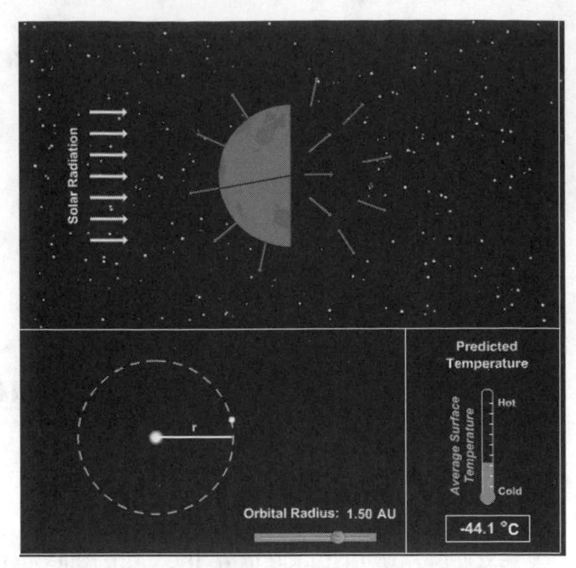

Figure 13-2 This tool from Lesson 2 estimates the temperature of a planet based on its distance from the Sun.

5. Explain why an average surface temperature predicted by the tool is not particularly meaningful for Earth or especially for Mercury.

6. Explain why the average surface temperature predicted by the tool for Venus is not accurate.

LESSON 3

Use the tool from Lesson 3 shown in Figure 13-3 to help answer the questions in this lesson.

7. Define *albedo* and explain why the value for Earth is an average.

8. Which planet has the highest albedo? Explain the reason for this.

9. How does albedo seem to affect the overall temperature of a planet?

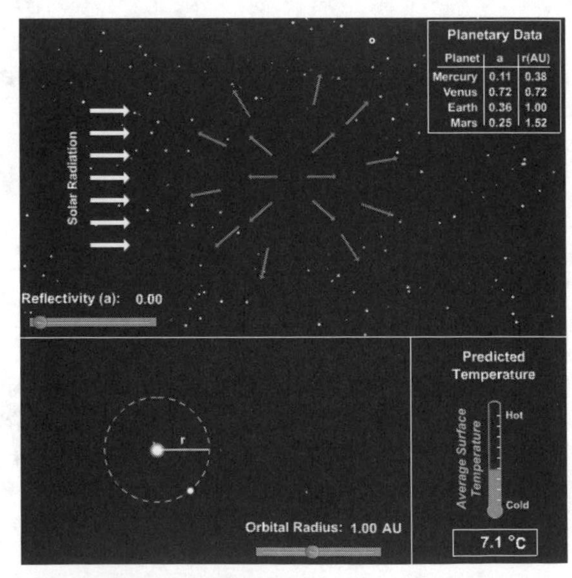

Figure 13-3 This tool from Lesson 3 estimates the temperature of a planet based on the planet's reflectivity (or albedo) and distance from the Sun.

LESSON 4

10. Name the three factors that affect a planet's surface temperature.

11. Which atmospheric gases affect the surface temperature of a planet (use the tool shown in Figure 13-4)?

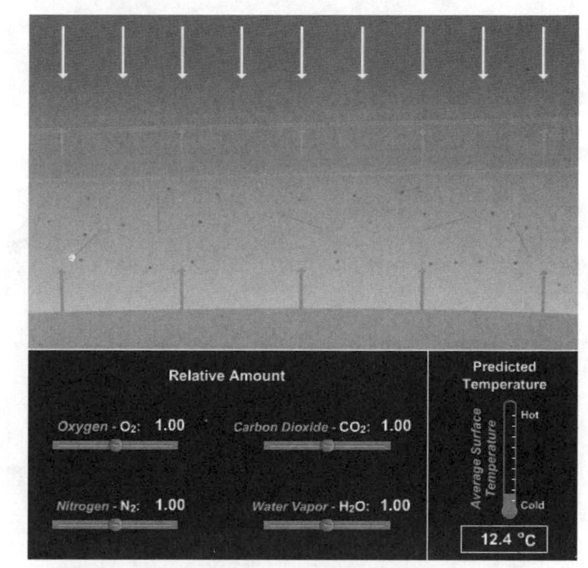

Figure 13-4 This tool from Lesson 4 demonstrates which gases affect the surface temperature of a planet.

12. Explain why Venus is so much hotter than the Earth.

LAB ACTIVITY Surface Temperature of Terrestrial Planets

The purpose of this activity is to learn what factors determine the surface temperature of planets.

For this activity, use the tool found in Lesson 4 (shown in Figure 13-5).

Use the version of the tool shown in Figure 13-5 (meant to answer questions about Venus in Lesson 4) to fill in the table below for Mercury and Venus. Then switch to the slightly different version (meant for Mars) in the same lesson to fill in the table for the Earth and Mars. The only difference is the range of relative amounts of CO_2 compared to the Earth.

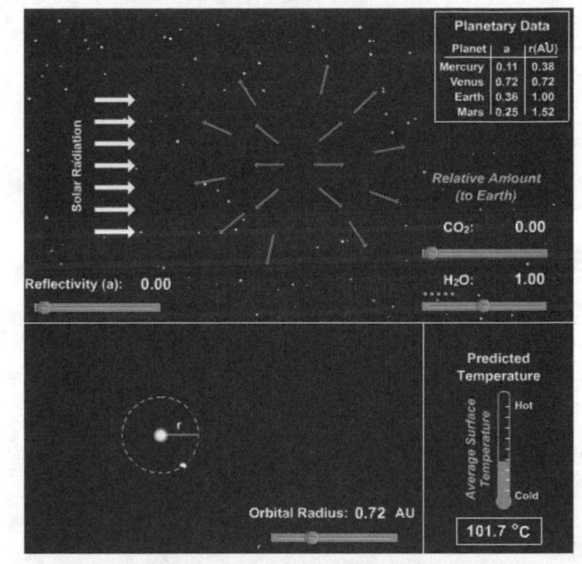

Figure 13-5 This tool from Lesson 4 estimates the temperature of a planet based on the planet's atmospheric composition, reflectivity (or albedo), and distance from the Sun. *Note: This tool is very similar to one in the Exercises section of the tutorial; be sure to use the one from Lesson 4.*

Planet	Albedo	r (AU)	CO_2	H_2O	Temperature (°C)	Realistic?
Mercury			0	0		
Venus			65000	0		
Earth			1	1		
Mars			13	0		

How would you expect surface temperature to vary with a planet's distance from the Sun? Does it vary in this way? Explain why or why not.

How would you expect surface temperature to vary with atmospheric content of Water Vapor and Carbon Dioxide (greenhouse gases)? Does it vary in this way? Explain why or why not.

How would you expect surface temperature to vary with albedo? Does it vary in this way? Explain why or why not.

14

Detecting Extrasolar Planets

Your goals in this tutorial are to understand:

- why visual detection of a planet orbiting another star is very difficult
- how the Doppler shift in a star's spectrum can be used to detect a planet and determine its mass and orbital radius
- how small but regular changes in the brightness of a star can be used to detect a planet in orbit

LESSON 1

1. Give two reasons why visual detection of planets orbiting other stars is close to impossible.

2. Which factors affect how much less luminous a planet will be than its star (see Figure 14-1)?

3. Do planets give off their own light? If not, how do they give off light?

4. As you move away from two objects, the apparent angle between them INCREASES | DECREASES | STAYS THE SAME *(circle one)*.

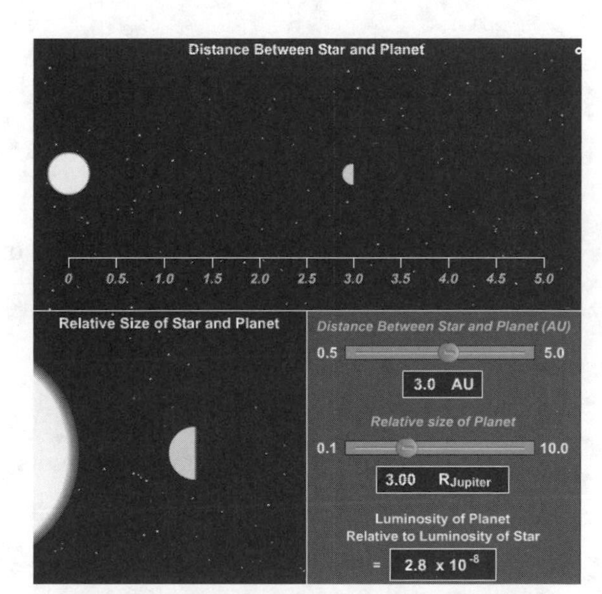

Figure 14-1 This tool from Lesson 1 determines the relative luminosity of a star and a planet based on the distance between the star and the planet and the size of the planet.

LESSON 2

5. Explain how the orbital period of an extrasolar planet is determined (see Figure 14-2).

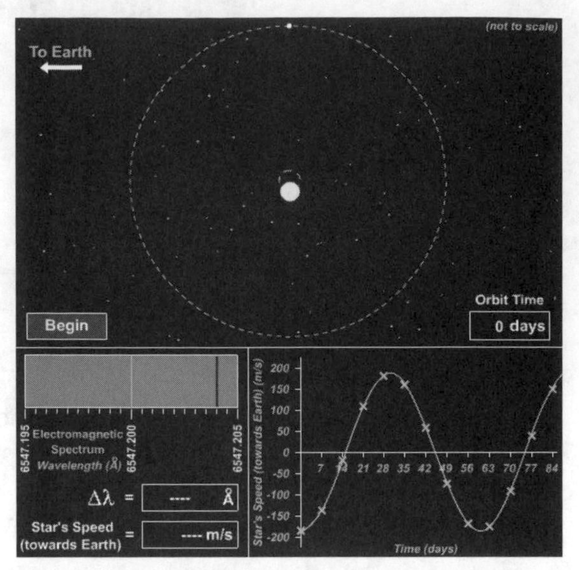

Figure 14-2 Use the tool in Lesson 2 to determine the period of an extrasolar planet.

6. How does the mass of a planet affect the Doppler shifts in its star's light?

7. How does the orbital radius of a planet affect the Doppler shifts in its star's light?

8. How is the mass of a planet related to its orbital period?

LESSON 3

9. Explain why our estimates of the masses of extrasolar planets are minimums.

10. Why are so many of the extrasolar planets that have been detected thus far as massive or more massive than Jupiter?

11. Why are so many of the extrasolar planets that have been detected thus far in orbits so close to their stars?

12. Which property of a star can be used to detect the presence of a planet and the planet's size? What change in this property suggests the presence of a planet, and why does it occur (see Figure 14-3)?

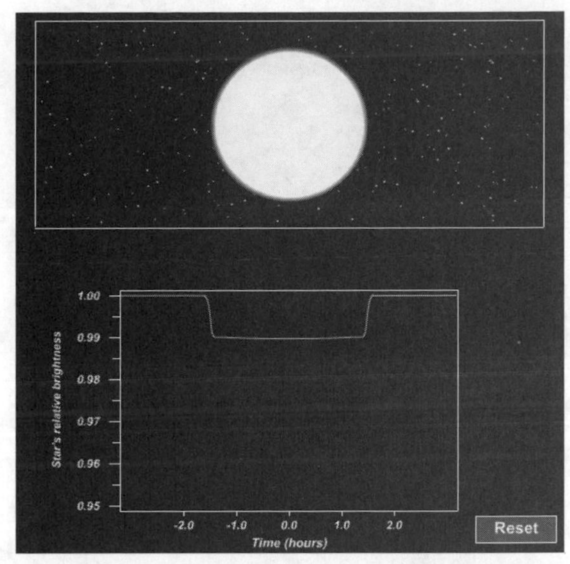

Figure 14-3 This tool shows what happens to a star's apparent luminosity when a planet transits in front of it.

LAB ACTIVITY Detecting Extrasolar Planets

The purpose of this activity is to learn how the properties of a planet affect the light observed from its star.

Use the tool from Lesson 2 (shown in Figure 14-4) to answer the questions.

Keeping the planet mass constant, vary the orbital radius to see how the star's speed varies. Then keep the orbital radius constant and vary the planet mass to see the effect on the star's speed. Now consider the four graphs below. Circle A on the graph showing the planet of larger mass in a closer orbit. Circle B on the graph showing larger mass in a farther orbit. Circle C on the graph showing a smaller mass in a closer orbit. Circle D on the graph showing a smaller mass in a farther orbit. Using the tool to reproduce each graph will help you make your choices.

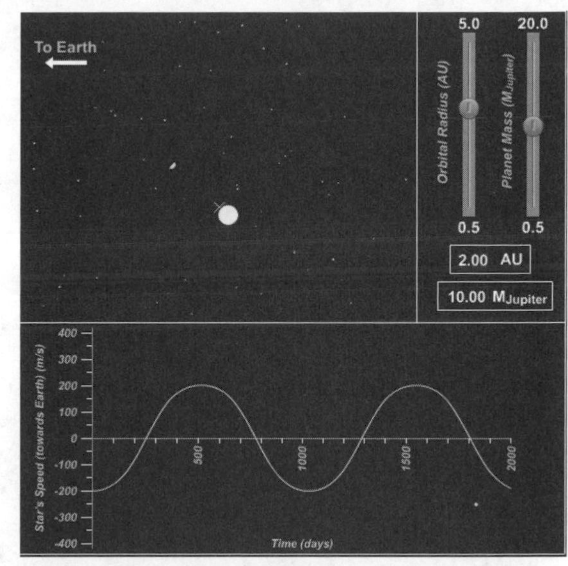

Figure 14-4 Use this tool from Lesson 2 to determine the effect of the orbital radius and mass of a planet on a star's orbital speed.

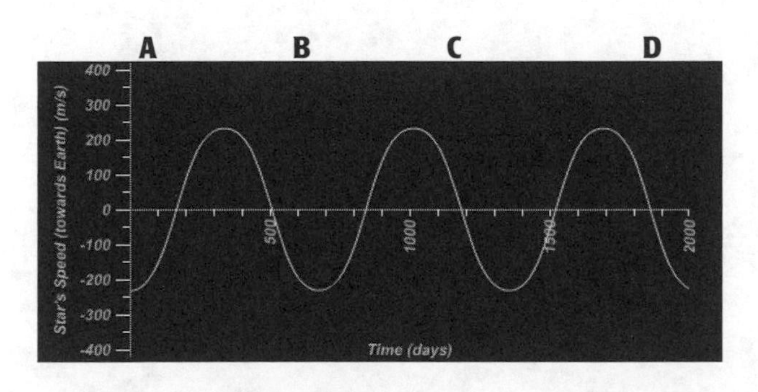

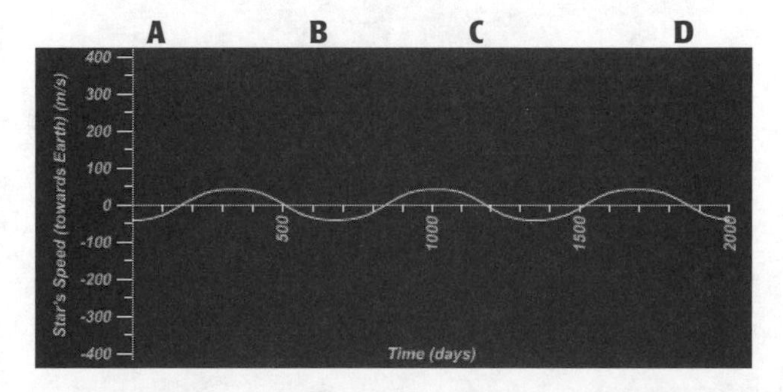

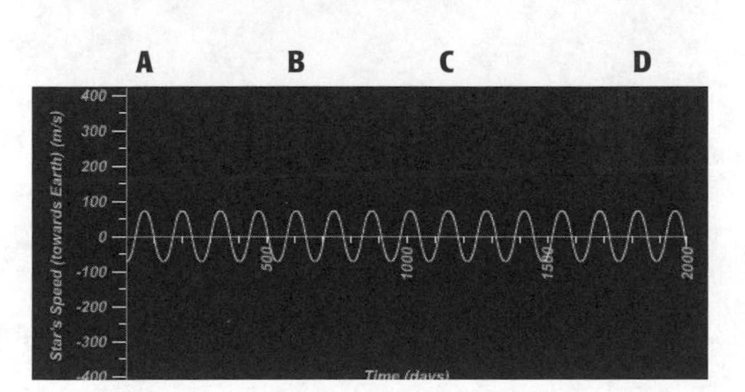

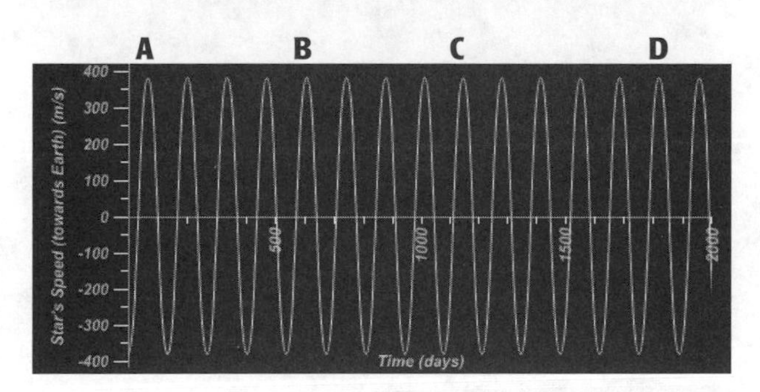

15
The Sun

The goals of this tutorial are to learn:

- the main difference between a star and a planet
- why and how nuclear fusion occurs in the core of the Sun

LESSON 1

1. List the four layers of the solar interior, in order, from the center to the surface of the Sun (see Figure 15-1).

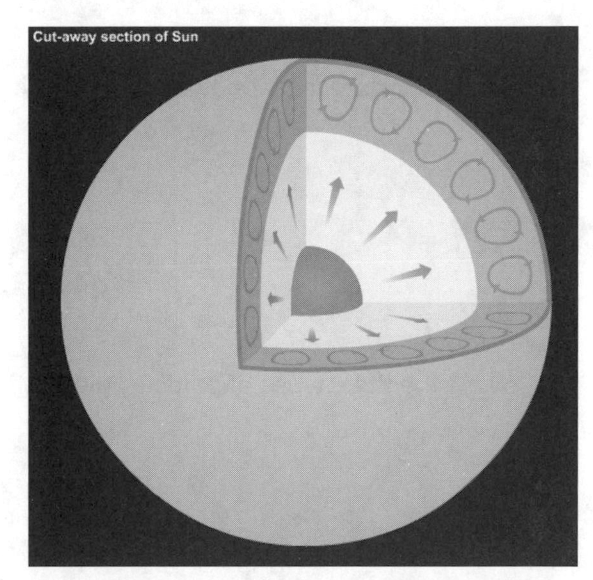

Figure 15-1 This diagram shows the layers of the Sun.

2. How does the temperature of the Sun vary from the center to the surface of the Sun? Explain the reason for this variation. How does the pressure vary? Explain the reason for this as well.

3. How does the energy of photons change as they travel from the core to the surface of the Sun? Explain the methods of energy transfer that occur and why they change as the photons travel from the core.

4. Name two processes that occur in the Sun that also occur in Earth's atmosphere. Explain how each process works.

LESSON 2

5. Explain the main difference between a star and a planet. What is the one property of a star or planet that is responsible for this difference?

6. Although the process of four hydrogen nuclei fusing into one helium nucleus is quite complex, what ultimately happens during this process that allows a star to give off light?

7. What important fact about the Sun can be estimated based on the rate of nuclear fusion occurring in the core?

8. Name the three types of particles given off during the proton-proton chain (see Figure 15-2).

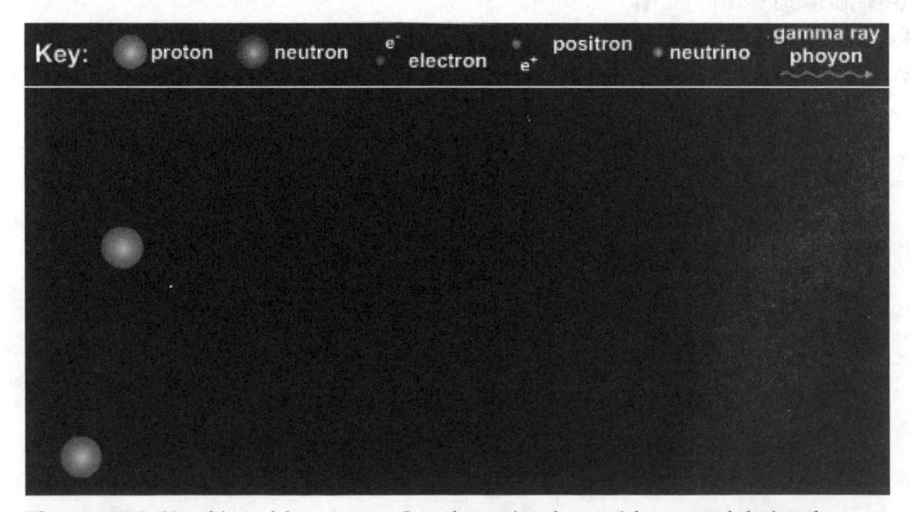

Figure 15-2 Use this tool from Lesson 2 to determine the particles created during the proton-proton chain.

LESSON 3

(Use Figure 15-3 to help answer questions 9 and 10.)

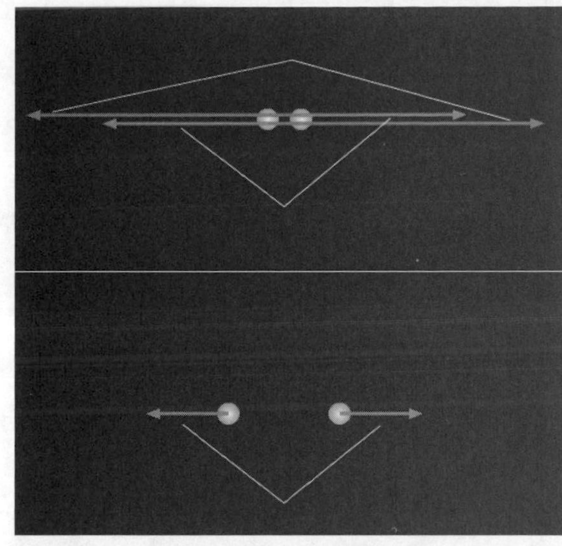

Figure 15-3 This tool allows you to compare electromagnetic and strong nuclear forces.

9. All protons carry positive charges. Explain how it is possible for protons to not repel one another and stay together in the nucleus of atoms heavier than hydrogen (such as helium).

10. Why do protons tend to repel one another? What is the key to getting them to "stick" together?

11. How can this key to "getting them to 'stick' together" in question 10 be achieved?

12. Based on your answer to question 11, explain why nuclear fusion occurs only in the core of a star.

LAB ACTIVITY The Sun

The purpose of this lab activity is to determine how the mass and luminosity of a star affect the star's lifetime.

Use the tool from Lesson 2 (shown in Figure 15-4) to answer the questions below.

Note: Although this tool is useful for showing the relationship between mass or luminosity and the lifetime of stars, it is important to know that, as you will learn in a coming tutorial, there is a relationship between mass and luminosity as well. More massive stars tend to be more luminous and less massive stars less luminous. There generally are not luminous low-mass stars and high-mass stars of low luminosity.

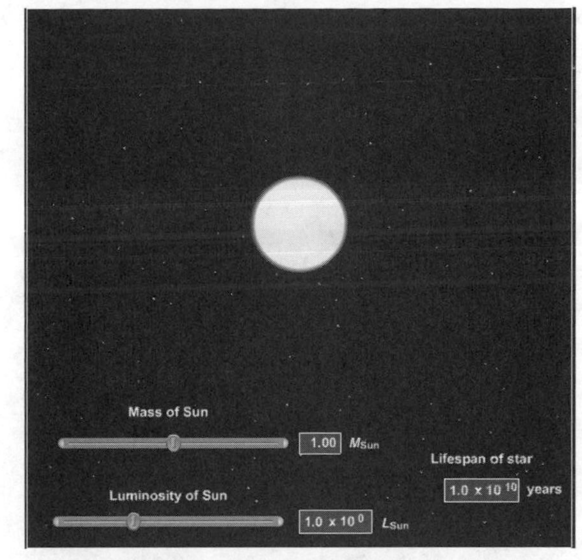

Figure 15-4 This tool from Lesson 2 determines the lifespan of stars with different masses and luminosities.

Vary the luminosity while leaving the mass at 1.00. How does varying the luminosity seem to affect the lifetime of a star?

Now vary the mass while leaving the luminosity at 1×10^0. How does varying the mass seem to affect the lifetime of a star?

Based on your above answers, predict how the lifetime of a star less massive and less luminous than our Sun would compare to the Sun's lifetime. Also predict how the lifetime of a more massive and more luminous star would compare to the Sun's.

Predictions:

A less massive and luminous star should live a *(circle one)* LONGER | SHORTER | SIMILAR LIFESPAN than the Sun.

A more massive and luminous star should live a *(circle one)* LONGER | SHORTER | SIMILAR LIFESPAN than the Sun.

Now use the tool to check your predictions.

Were your predictions correct? Did the results of your tests surprise you? Explain your answers.

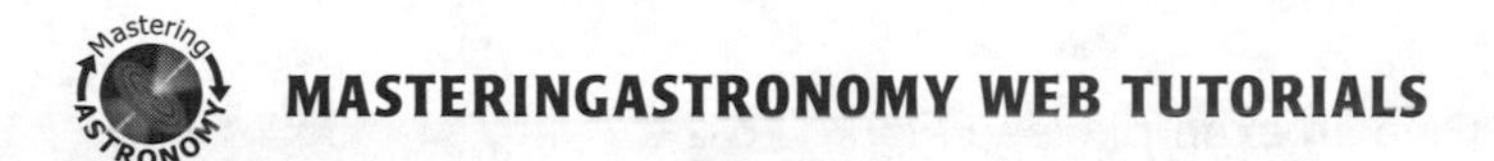

16
Measuring Cosmic Distances

Your goals in this tutorial are to:

- list the most accurate methods of determining distances for objects within our solar system, nearby stars, nearby galaxies, and distant galaxies
- discuss how the parallax angle of a star depends on its distance from Earth
- explain how an object's distance can be determined if its luminosity and apparent brightness are known
- list four standard candles and how their luminosities are found

LESSON 1

1. Give two reasons that radar (rather than sonar) is used to determine distances in the solar system. Why is sonar better on the Earth?

2. Radio waves travel at the speed of SOUND I LIGHT *(circle one)*.

3. When at opposition, Jupiter is about five times farther from the Sun than Earth. If radar takes eight minutes to travel from the Earth to the Sun, how long would it take radar to travel from the Earth to Jupiter and back when Jupiter is at opposition. (see Figure 16-1)?

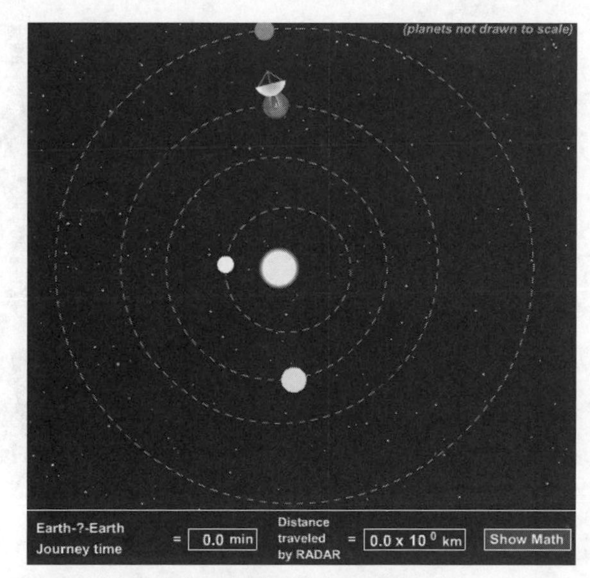

Figure 16-1 When Jupiter is at opposition, its configuration with Earth is the same as Mars's except that Jupiter is farther away.

LESSON 2

4. What is parallax? Give an example (see Figure 16-2).

5. Explain why radar cannot be used to measure the distance to stars.

6. Explain why parallax does not work for measuring the distance to all stars.

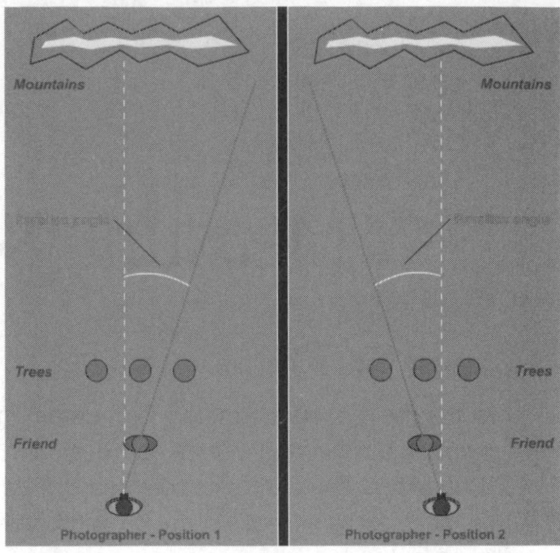

Figure 16-2 This diagram shows one example of parallax.

LESSON 3

7. What makes it possible for a streetlight to appear so much brighter than a star (see Figure 16-3)?

8. How does the apparent brightness of an object vary with an observer's distance from it?

9. Explain what is meant by a *standard candle*.

Figure 16-3 A candle can appear as bright as a lighthouse.

LESSON 4

10. List the following methods for determining distance in order of their ability to measure increasing distances. That is, the list should start with the method for measuring the distances to the closest objects. Then, for each method, list an object whose distance would be measured using that method.

    Fitting; Hubble Law; Parallax; Radar; Tully-Fisher Relationship

11. Each method used to measure farther distances depends on results from the previous method. Errors in the first method thus are magnified with each successive method used. Explain why larger errors in the distance to farther objects are not as significant a problem as errors in the distance to closer objects.

12. The Tully-Fisher relationship is a correlation between which two properties of a galaxy (see Figure 16-4)? How is it used to measure the distance to a galaxy?

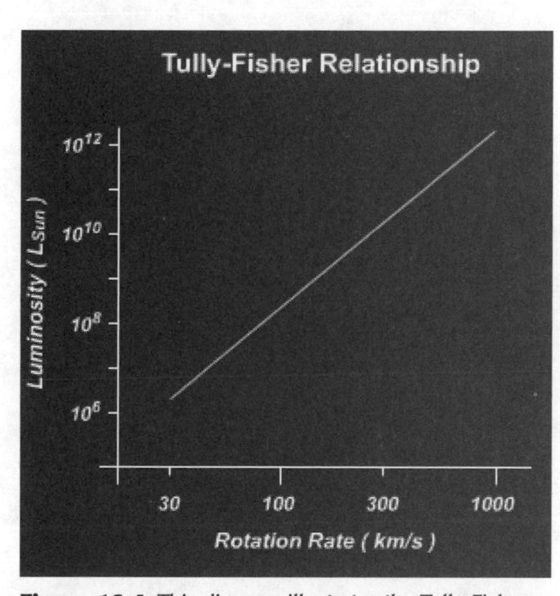

Figure 16-4 This diagram illustrates the Tully-Fisher relationship.

LAB ACTIVITY Measuring Cosmic Distances

The purpose of this activity is to determine how the apparent brightness of an object changes with one's distance from it.

Use the tool from Lesson 3 (shown in Figure 16-5) to complete the table and answer the questions.

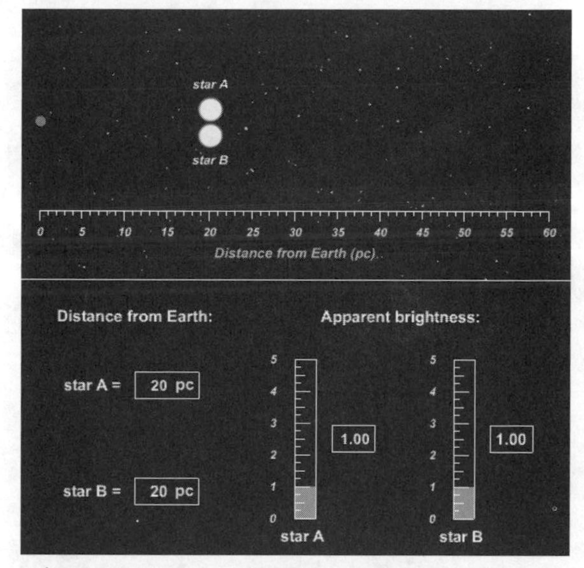

Figure 16-5 This tool from Lesson 3 compares the apparent brightness to the distance of a star.

Star A Distance (pc)	Star B Distance (pc)	Distance Ratio A/B	Star A Apparent Brightness	Star B Apparent Brightness	Brightness Ratio A/B
10	20			1	
20	20			1	
40	20			1	
60	20			1	

When Star A is half as far away as Star B, Star A appears ________________ times as bright.

When Star A is twice as far away as Star B, Star A appears ________________ times as bright.

When Star A is three times as far away as Star B, Star A appears ________________ times as bright.

Which best describes the relationship between brightness (B) and distance (d)? *(circle one)*

$B = d$ $B = d^2$ $B = 1/d$ $B = 1/d^2$

Plot a graph on the axes in Figure 16–6 of the distance ratio versus the brightness ratio.

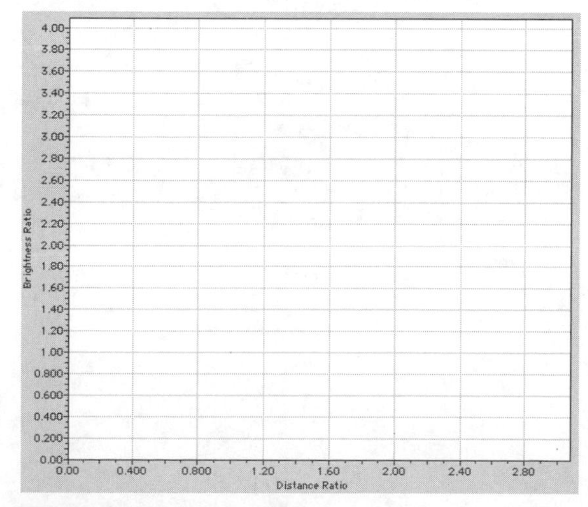

Figure 16-6 Distance ratio.

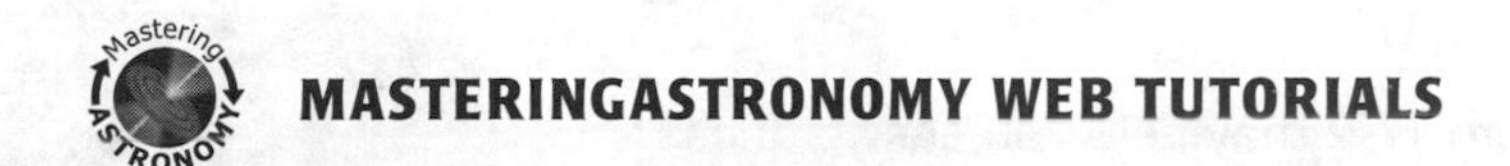

17

The Hertzsprung-Russell Diagram

Your goals in this tutorial are to:

- generate an H-R Diagram using the luminosities and temperatures of several stars
- explain the spectral classification scheme for stars
- use an H-R Diagram to determine a star's radius
- explain how the mass of a star determines its location on the H-R Diagram

LESSON 1

1. If one star appears brighter than another, can you be sure that it is actually brighter? Why or why not?

Use Figure 17-1 to answer questions 2–4.

2. List the four most common star colors (red, white, blue, and yellow) in order from coolest to hottest.

3. Stars of each of the four colors from question 2 are of which spectral type (OBAFGKM)?

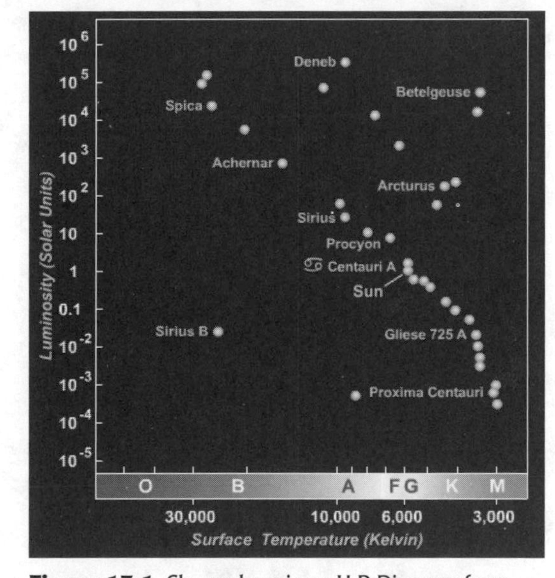

Figure 17-1 Shown here is an H-R Diagram from Lesson 1. Use this tool to answer questions 2–4.

4. Match the properties of a star on the left with where it will be found on the H-R Diagram.

Hot	Low
Cool	Right
Bright	Left
Dim	High

LESSON 2

The tool shown in Figure 17-2 may help you answer these questions.

5. What colors are the brightest stars? The dimmest stars?

6. How does brightness vary with the size of a star?

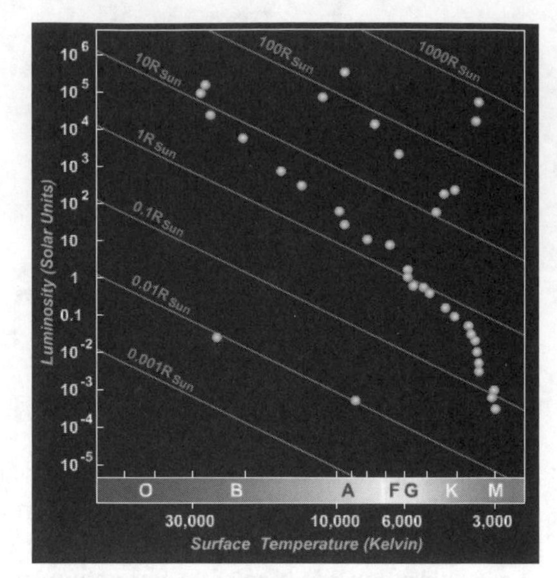

Figure 17-2 This illustration shows the lines of equal radii on an H-R Diagram.

7. Which two factors determine the brightness of a star?

8. How does one star compare in size to another if it is directly to the left of the other on the H-R Diagram?

9. How does one star compare in size to another if it is directly above the other on the H-R Diagram?

LESSON 3

Use the tool shown in Figure 17-3 to answer the questions in this lesson.

10. Which main-sequence star is more massive, a brighter or a dimmer one? A hotter or a cooler one?

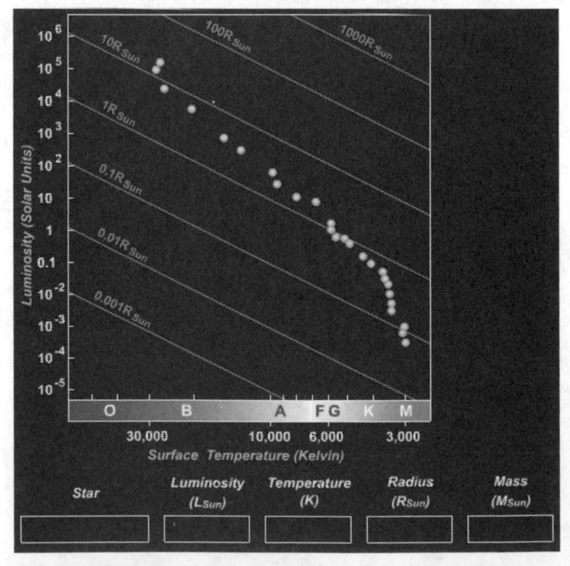

Figure 17-3 This H-R Diagram from Lesson 3 includes the radii and masses of stars.

11. List the four main-sequence star colors (red, white, blue, and yellow) in order from most to least massive.

12. A certain red giant star is bigger and brighter than a certain blue giant star. Can you tell for sure which is more massive? Explain your answer.

LAB ACTIVITY The Hertzsprung-Russell Diagram

The purpose of this activity is to use the luminosity and temperature of stars on the H-R Diagram to name them according to color and size.

Use the tool from Lesson 1 (shown in Figure 17-4) to complete the table. In the last column, name each star in terms of its type and its color and brightness—i.e., bright and cool = red giant, dim and hot = white dwarf. Every star on the table must be named in this way.

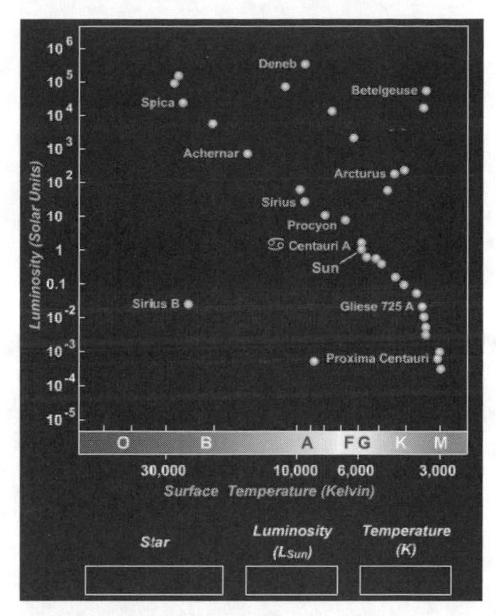

Figure 17-4 Use this H-R Diagram from Lesson 1 for the lab activity.

Star Name	Luminosity	Temperature	Star Type
Achemar			
Arcturus			
Betelguese			red giant
Centauri A			
Deneb			
Gliese 725 A			
Procyon			
Proxima Centauri			red dwarf
Sirius A			
Sirius B			
Sun			
Spica			

Considering stars only on the main sequence and recalling the answers to the questions from Lesson 3, which main-sequence stars from your table are the most massive? What kind of stars are they?

Which main-sequence stars from your table are the least massive? What kind of stars are they?

What kind of star is our Sun? Is it among the most or least massive of main-sequence stars?

MASTERINGASTRONOMY WEB TUTORIALS

18

Stellar Evolution

Your goals in this tutorial are to:

- understand the relationship between the mass and lifetime of a main-sequence star
- discuss the basic stages of stellar evolution for low- and high-mass stars
- explain why the H-R Diagram of a star cluster changes with time
- determine the age of the cluster from its H-R Diagram

LESSON 1

1. What event that occurs inside a star is a signal that it is beginning to die?

Use the H-R Diagram tool shown in Figure 18-1 to help you answer questions 2–4.

2. How is the length of a star's life related to its mass? Which will live longer, a brighter or a dimmer main-sequence star? Explain your answer.

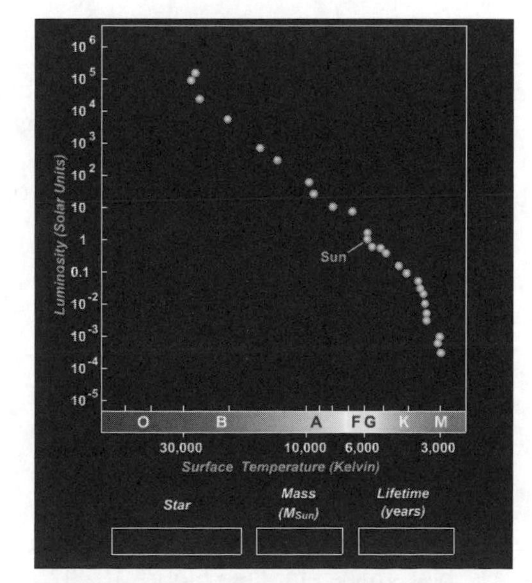

Figure 18-1 This H-R Diagram from Lesson 1 is a plot of the temperature versus luminosity of stars.

3. A blue giant star is MORE | LESS *(circle one)* massive than our Sun. It is also BRIGHTER | DIMMER *(circle one)* and HOTTER | COOLER *(circle one)* than our Sun but will live a LONGER | SHORTER *(circle one)* life.

4. A red dwarf star is MORE | LESS *(circle one)* massive than our Sun. It is also BRIGHTER | DIMMER *(circle one)* and HOTTER | COOLER *(circle one)* than our Sun but will live a LONGER | SHORTER *(circle one)* life.

LESSON 2

The illustration in Figure 18-2 may be helpful with the questions in this lesson.

5. After main sequence, what is the next major stage in the life of a star like our Sun?

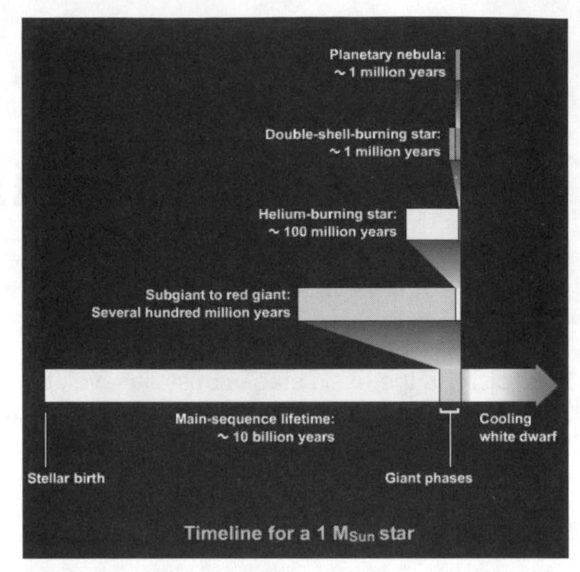

Figure 18-2 This illustration shows the stage of the life of a solar-mass star.

6. During this next stage, the star will become BRIGHTER | DIMMER *(circle one)*, HOTTER | COOLER *(circle one)*, and LARGER | SMALLER *(circle one)* than it is now.

7. In reaching this stage, which direction(s) will the star move on the H-R Diagram?

8. What is the next major stage after the one identified in question 5?

9. During this final stage, the star will become BRIGHTER | DIMMER *(circle one)*, HOTTER | COOLER *(circle one)*, and LARGER | SMALLER *(circle one)* than it is now.

10. In reaching this stage, which direction(s) will this star move on the H-R Diagram?

LESSON 3

11. What event occurs at the end of the life of a high-mass star (see Figure 18-3)?

Figure 18-3 Shown here is a supernova remnant.

LESSON 4

12. In the first animation Lesson 4 (see Figure 18-4a), list the order in which you saw the differently colored stars in the cluster disappear. Explain the reason that they disappeared in this order.

13. In the second animation Lesson 4 (see Figure 18-4b), the *(circle one)* BRIGHTEST | DIMMEST stars on the main sequence disappeared first. These stars are also the *(circle one)* MOST | LEAST MASSIVE. Explain the reason for the order in which the stars disappeared.

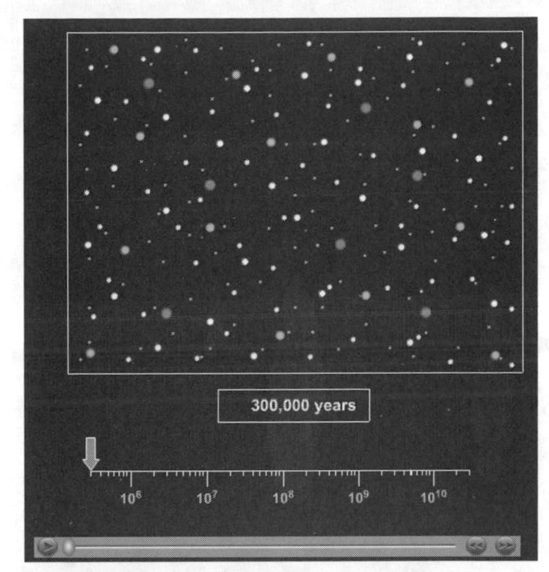

Figure 18-4a Stars of different masses have lifetimes of different lengths, changing the composition of a star cluster over time.

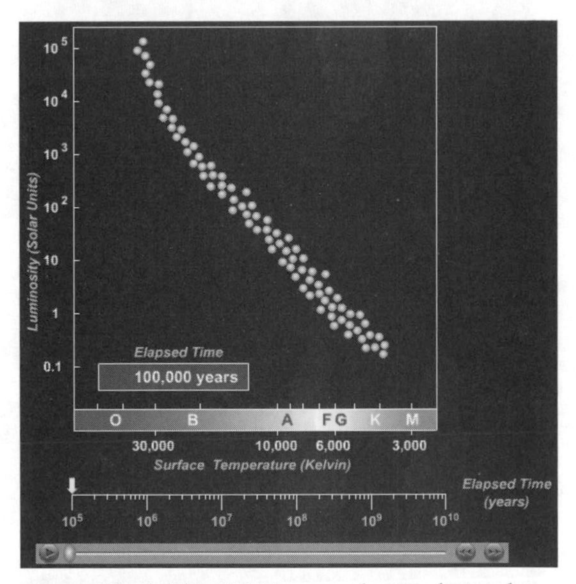

Figure 18-4b This H-R Diagram of a star cluster also changes over time due to the differing lifetimes of stars with different masses.

LAB ACTIVITY Stellar Evolution

The purpose of this activity is to see how the mass of a star is related to its lifetime and luminosity.

Use the tool in Lesson 1 (shown in Figure 18-5) to determine the time each star in the table will take to change all the hydrogen in its core into helium.

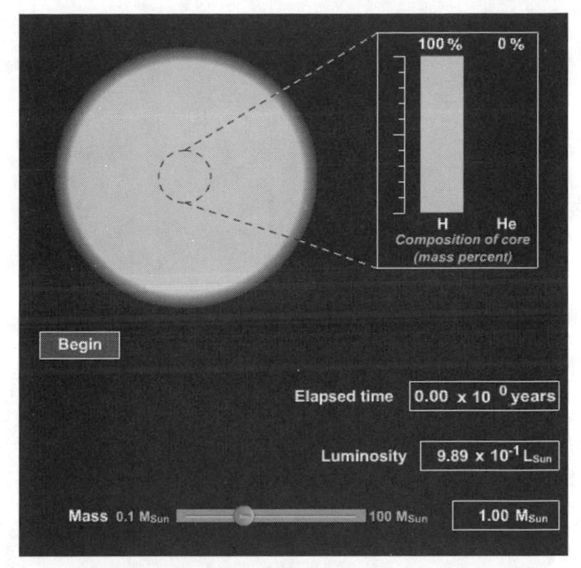

Figure 18-5 This tool from Lesson 1 determines the lifetime and luminosity of stars of different masses.

Mass (M_{Sun})	Elapsed Time (years)	Luminosity (L_{Sun})
$0.1 = 10^{-1}$		
$1 = 10^{0}$		
$10 = 10^{1}$		
$100 = 10^{2}$		

More massive stars live LONGER | SHORTER *(circle one)* lives than less massive ones. Brighter stars live LONGER | SHORTER *(circle one)* lives than dimmer ones.

Can you think of a reason for your answers?

Plot your data from the table on the graphs on the next page.

Note: Plot only the exponents from your data.

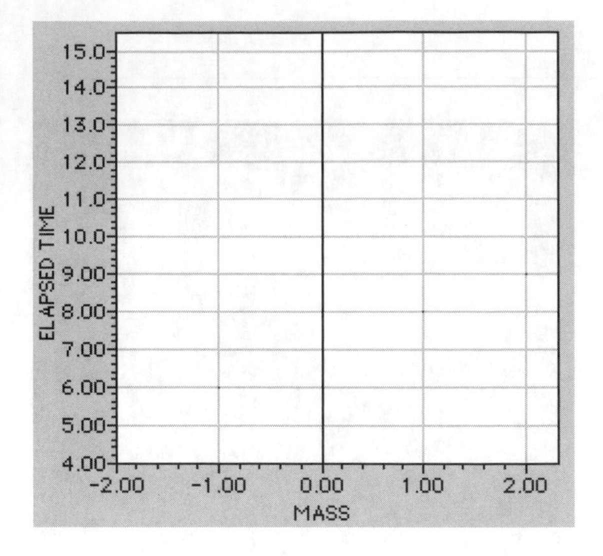

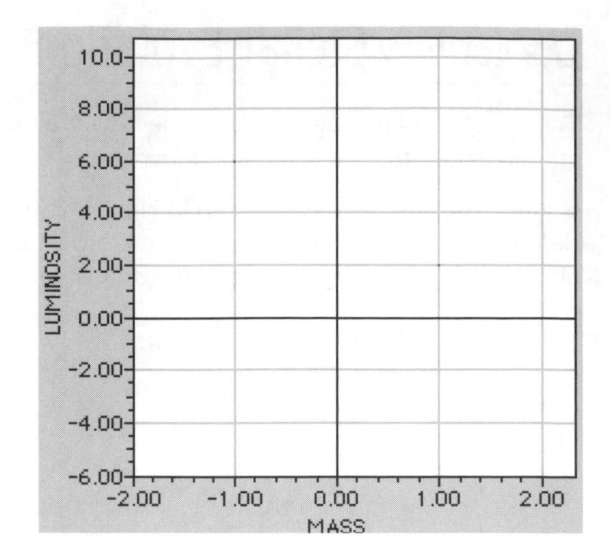

More massive stars are BRIGHTER | DIMMER *(circle one)*. This is known as the *Mass-Luminosity Relationship*.

MASTERINGASTRONOMY WEB TUTORIALS

19
Black Holes

The goals of this tutorial are to:

- describe what a black hole is
- discuss the significance of the event horizon
- discuss the evidence that black holes exist

LESSON 1

1. Define *escape velocity*. Define a black hole in terms of escape velocity.

2. How does the mass of an object affect escape velocity?

3. How does the size (radius) of an object affect escape velocity?

4. Explain why the escape velocity of a heavyweight star increases when it collapses.

5. What is an event horizon?

6. What is the Schwarzschild radius?

LESSON 2

7. Why can't we see a black hole?

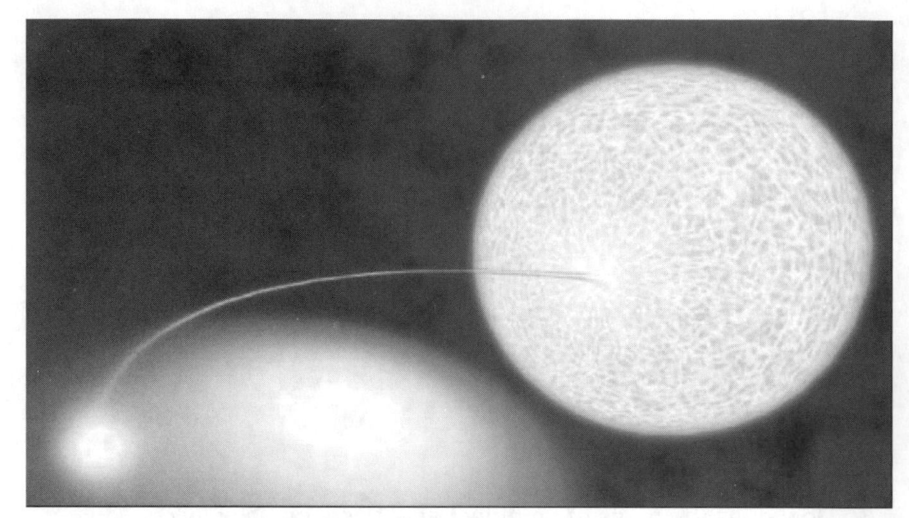

Figure 19-1 Black holes can be part of binary systems.

8. If we can't see a black hole, how can we detect one (see Figure 19-1)?

9. What is the name given to material orbiting the event horizon of a black hole?

10. What kind of energy does material orbiting a black hole emit?

11. What happens to the size of a black hole when material crosses the event horizon?

12. Where is a good place to look for a supermassive black hole (see Figure 19-2)?

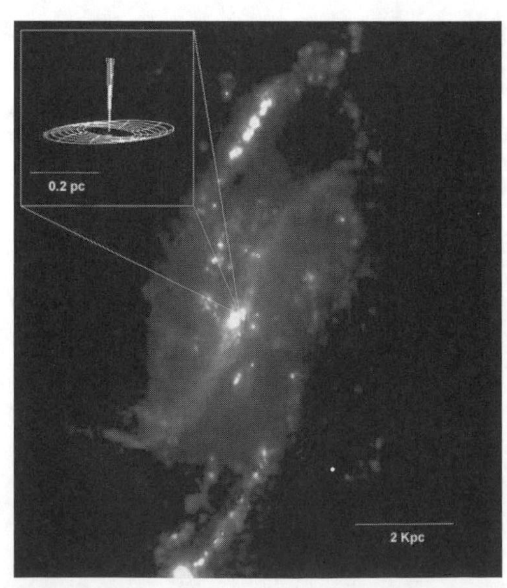

Figure 19-2 Supermassive black holes are believed to be at the center of many galaxies including our own Milky Way.

LAB ACTIVITY Black Holes

The purpose of this activity is to determine the relationship between the mass and size of a black hole.

Use the tool from Lesson 1 (shown in Figure 19-3) to determine the Schwarzschild radius for black holes of different masses. Record the masses on the table and answer the question.

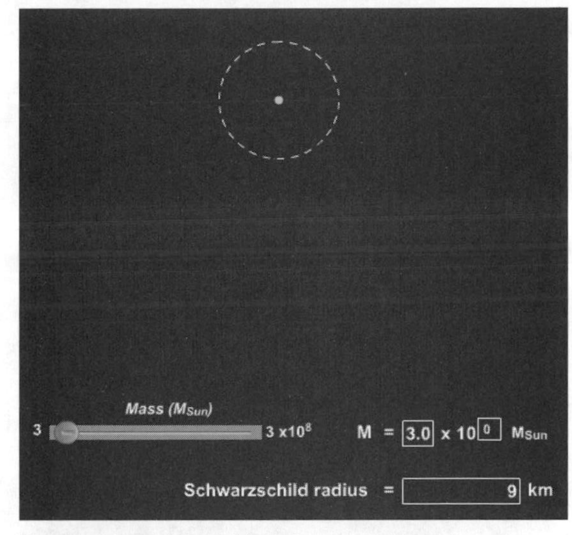

Figure 19-3 This tool from Lesson 1 determines the Schwarzschild radius of a black hole as a function of its mass.

Mass (M_{Sun}) = M	Schwarzschild radius (km) = R	R/M
3	9	
5		
10		
20		
30		
50		
60		
90		
100		

Average your values for R divided by M (R/M) and use the value to complete the formula:

$$\mathbf{R} = \_\_\_\_\_\_\_\_\_\_\_\_ \times \mathbf{M}$$

Does the ratio of the mass to the radius of a black hole appear to change with mass or with radius?

20

Detecting Dark Matter in Spiral Galaxies

Your goals in this tutorial are to:

- explain what a rotation curve is
- sketch the rotation curves for planetary systems and spiral galaxies
- describe how the shapes of the rotation curves depend on the distribution of mass in the gravitational system
- explain how the curves for galaxies reveal the presence of dark matter

LESSON 1

1. On a merry-go-round, a person closer to the center moves FASTER | SLOWER *(circle one)* than a person on the edge.
2. In the solar system, planets closer to the Sun orbit FASTER | SLOWER *(circle one)* than planets farther from the Sun.
3. Would changing the speed of the merry-go-round change the relationship between the speed of the person on the edge and the speed of the person toward the center?

4. Would changing the mass of the Sun change the relationship between the speed of the planets in the closest orbits and the speeds of the planets in the farthest orbits?
5. Near its center, the rotation curve of a real galaxy is more like a MERRY-GO-ROUND | SOLAR SYSTEM | NEITHER *(circle one).*
6. At the edges, the rotation curve of a real galaxy is more like a MERRY-GO-ROUND | SOLAR SYSTEM | NEITHER *(circle one).*

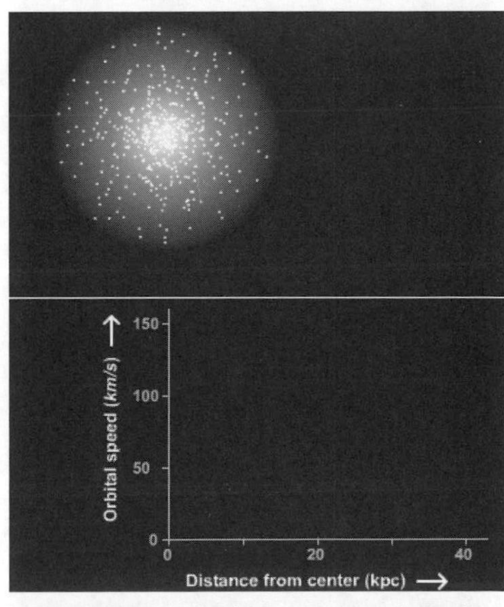

Figure 20-1 This tool from Lesson 1 determines the rotation curve for a galaxy. Use this tool to answer questions 5 and 6.

LESSON 2

7. Orbital speed is DIRECTLY | INVERSELY *(circle one)* proportional to the mass of the central object.

8. Orbital speed is DIRECTLY | INVERSELY *(circle one)* proportional to the distance from the central object.
9. What can we infer about the distribution of mass in our galaxy by observing the orbital speeds of stars as they get farther from the center of our galaxy? Explain your answer (see Figure 20-2).

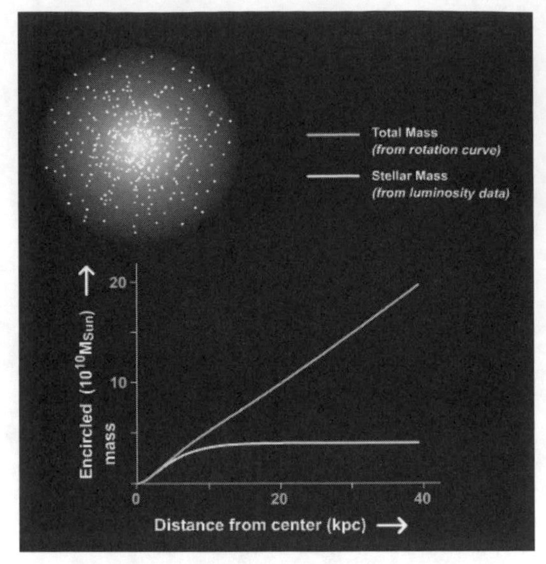

Figure 20-2 This diagram shows the plots of luminosity and mass distribution in a galaxy.

LESSON 3

10. Explain how stellar rotation curves for our galaxy suggest the existence of dark matter.

11. The amount of dark matter seems to INCREASE | DECREASE | REMAIN THE SAME *(circle one)* as you move away from the center of a galaxy (see Figure 20-3).
12. If dark matter could be anything except the stars in our galaxy, what do you think it might be?

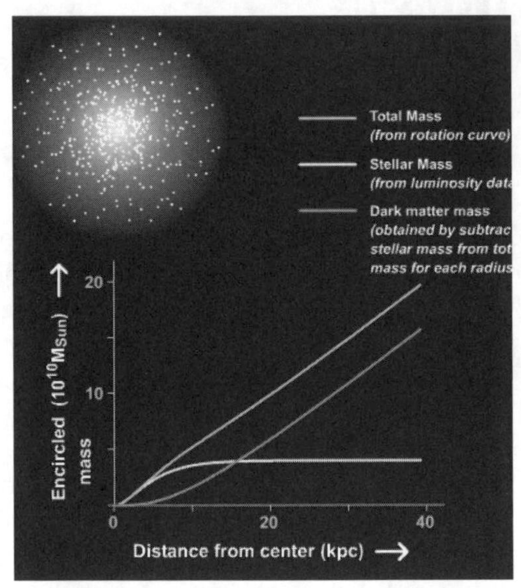

Figure 20-3 Shown here are the distributions of total mass, stellar mass, and mass of dark matter in our galaxy.

LAB ACTIVITY Detecting Dark Matter in Spiral Galaxies

The purpose of this activity is to see how the rotation curves of a solid object, a solar system, and a galaxy differ.

On the three axes below, sketch the rotation curves for a merry-go-round, a solar system, and a galaxy. Review Lesson 1 for help. Ignore the numbers on the axes; just draw the shapes of the graphs.

Merry-go-round

Reason for Shape

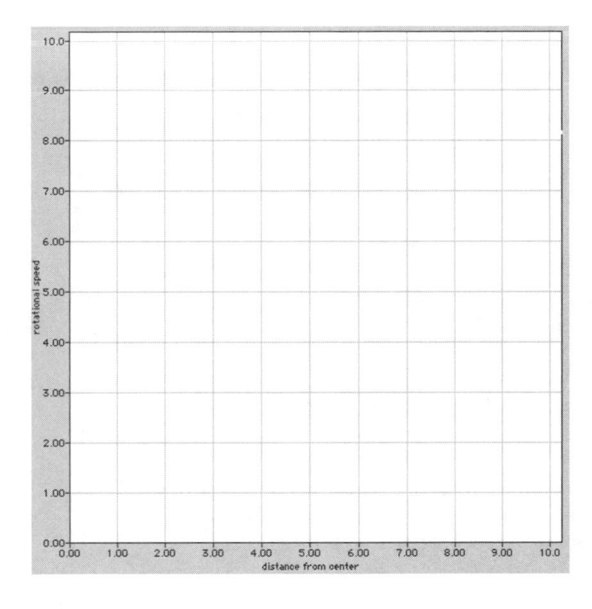

Solar system

Reason for Shape

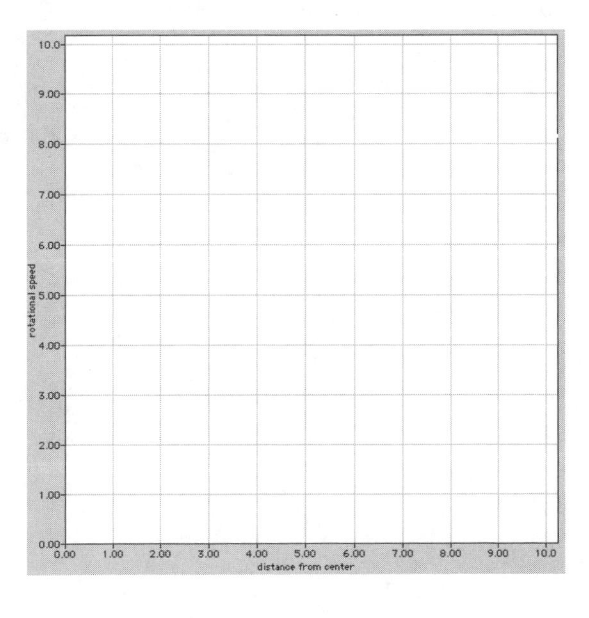

Galaxy

Reason for Shape

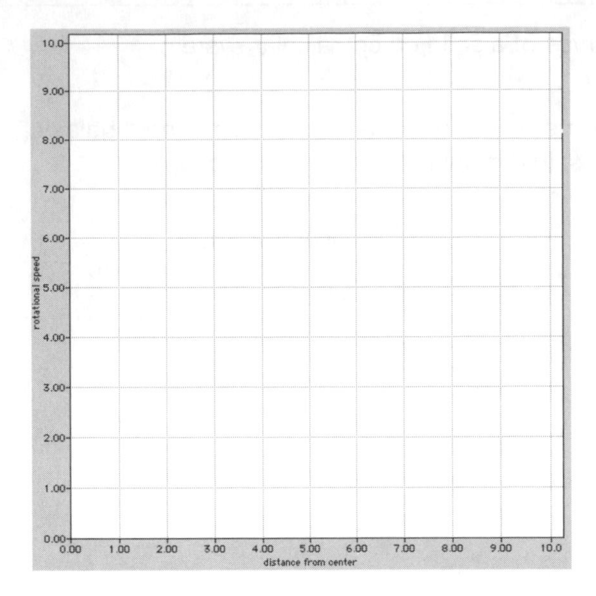

If you doubled your distance from the center of a merry-go-round, your orbital speed would *(circle one)*

STAY ABOUT THE SAME | INCREASE but LESS THAN DOUBLE | DOUBLE | MORE THAN DOUBLE.

If you doubled your distance from the center of the solar system, your orbital speed would *(circle one)*

STAY ABOUT THE SAME | INCREASE | DECREASE.

If you started a good distance from the center and then doubled your distance from the center of a galaxy, your orbital speed would *(circle one)*

STAY ABOUT THE SAME | INCREASE BUT LESS THAN DOUBLE | DOUBLE | MORE THAN DOUBLE.

MASTERINGASTRONOMY WEB TUTORIALS

21
Hubble's Law

Your goals in this tutorial are to:

- explain Hubble's law
- discuss how it implies that the universe is expanding
- determine Hubble's constant
- discuss how the age of the universe depends on it

LESSON 1

1. All distant galaxies (that is, those not in our own local cluster) show a BLUESHIFT | REDSHIFT *(circle one).*
2. This shift suggests that they are moving TOWARD US | AWAY FROM US | NEITHER *(circle one).*
3. To what does the amount of shift seem to be proportional (see Figure 21-1)?

4. Is this shift a Doppler shift? Explain your answer.

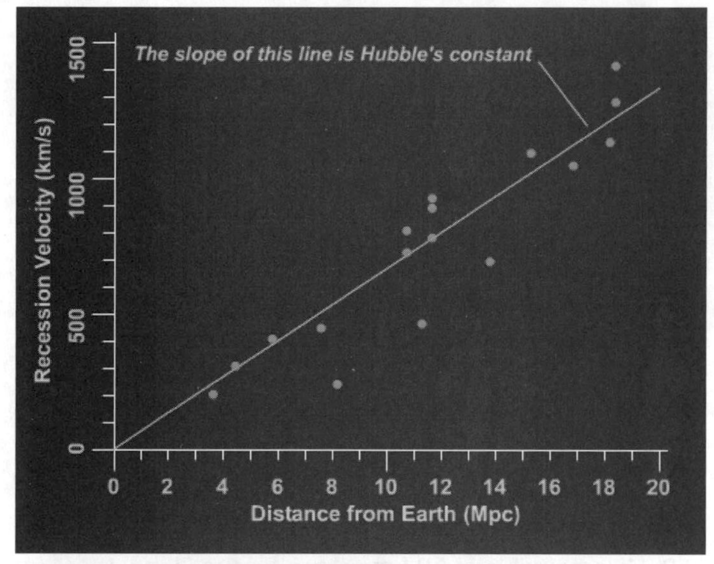

Figure 21-1 This diagram shows Hubble's law through a plot of the recessional velocity of galaxies versus their distance from Earth. The slope of the line is Hubble's constant.

LESSON 2

5. What does the observed shift in the light from distant galaxies suggest about the universe as a whole (see Figure 21-2)?

6. What does the answer to the previous question and the first two questions in this tutorial suggest about our position in the universe?

7. Does the answer to question 6 above seem likely? Explain why it is not actually true.

8. Let's say one galaxy is 4 Mpc away from us and another is 16 Mpc away. If the nearer one moves 1 Mpc, how far will the more distant galaxy move in the same amount of time?

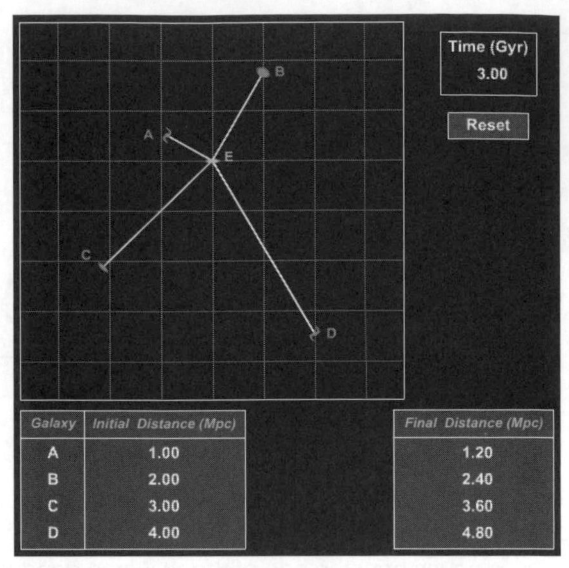

Figure 21-2 The distance to galaxies changes with time, as the farther galaxies move proportionally farther away.

9. How far away will each of the galaxies in question 8 be after that amount of time has passed?

LESSON 3

Use the tool shown in Figure 21-3 for these questions.

10. Using a value of 50 km/s/Mpc for the Hubble constant, what is the age of the universe?

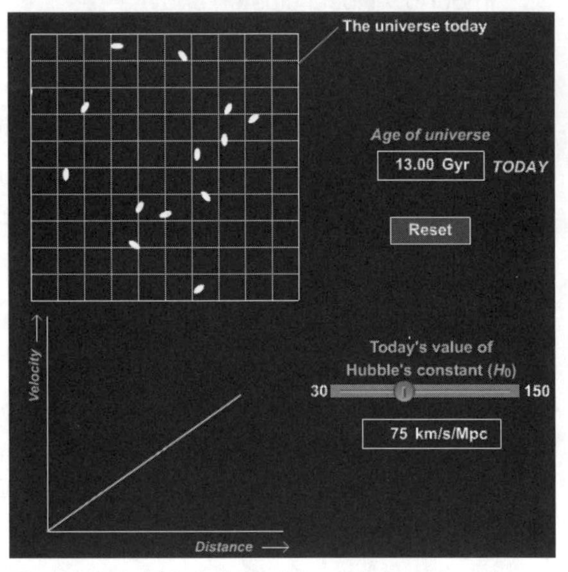

Figure 21-3 This tool from Lesson 3 determines the age of the universe for varying values of Hubble's constant.

11. How would a higher value for the Hubble constant affect the calculation of the age of the universe? A lower value?

12. If the value of the Hubble constant were 50 km/s/Mpc, what would be the farthest distance away at which we could see an object? Explain your answer. Is it possible that there are objects farther away that we cannot see? Explain your answer.

LAB ACTIVITY Hubble's Law

The purpose of this activity is to show how we can use Hubble's law to determine the distance to galaxies, their recessional velocities, and the age of the universe.

Use the graph in Figure 21-4 to determine the distance (in Mpc) to a galaxy that has a recessional velocity of about 400 km/s.

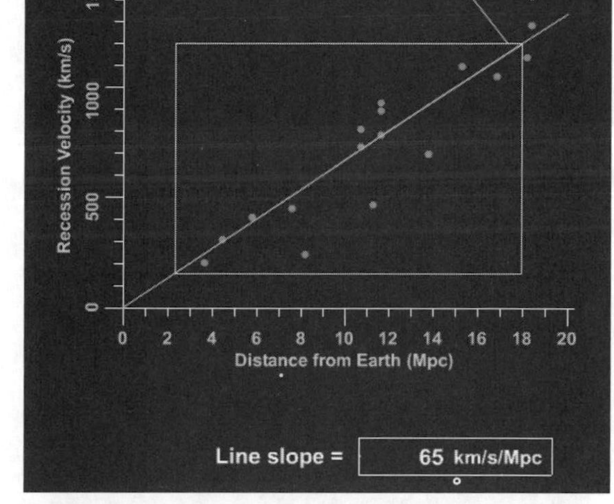

Figure 21-4 This tool from Lesson 1 determines Hubble's constant, the slope of a Hubble's law plot.

Use the graph in Figure 21-4 to determine the recessional velocity (in km/s) of a galaxy that is about 14 Mpc away.

The age of the universe T, in billions of years, can be approximated by the following expression:

$$T = 1000/H$$

where H is Hubble's constant.

Use the value of the Hubble constant in the graph in Figure 21-4 to determine the age of the universe.

T = ______________ billion years

22

Fate of the Universe

Your goals in this tutorial are to:

- discuss why the expansion rate of the universe may change with time
- describe the significance of the critical mass density of the universe
- predict the fate of the universe based on its value of critical mass density

LESSON 1

See the tool in Figure 22-1 for questions 1 and 2.

1. What are the two possible fates of a ball thrown from a planet?

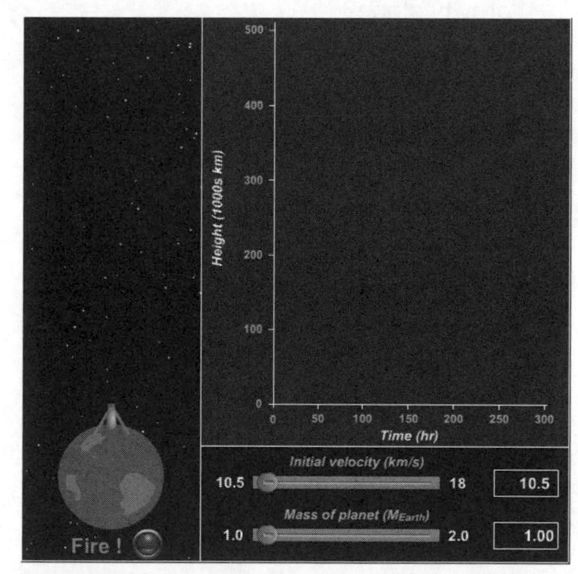

Figure 22-1 This tool from Lesson 1 determines the height attained by a cannon ball as a function of the initial velocity of its launch and the mass of the planet.

2. On what two factors does the fate of the ball in question 1 depend?

3. On what factors does the ultimate fate of the universe depend?

4. What are the possible fates of the universe?

5. Based on current estimates of the mass density of the universe, which of the fates in question 4 seems most likely?

LESSON 2

See the tool shown in Figure 22-2 for these questions.

6. Gravity tends to SLOW DOWN | SPEED UP *(circle one)* the expansion of the universe.
7. Dark energy tends to SLOW DOWN | SPEED UP *(circle one)* the expansion of the universe.
8. How could the presence of dark energy change the fate of the universe, compared to considering gravity alone?

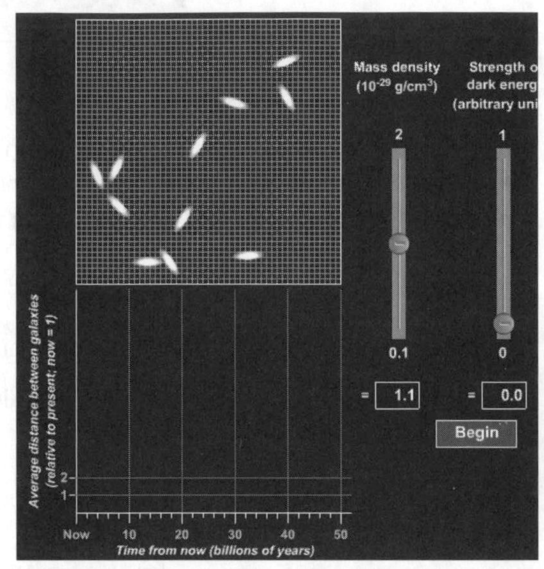

Figure 22-2 This tool from Lesson 2 determines the effects of gravity and dark energy on the fate of the universe.

LESSON 3

9. The universe would be YOUNGER | OLDER *(circle one)* than presently thought if dark energy did *not* exist (see Figure 22-3).
10. Current data seem to show that dark energy DOES | DOES NOT *(circle one)* exist.
11. Based on current data, what is the age of the universe?
12. Based on current data, what is the ultimate fate of the universe?

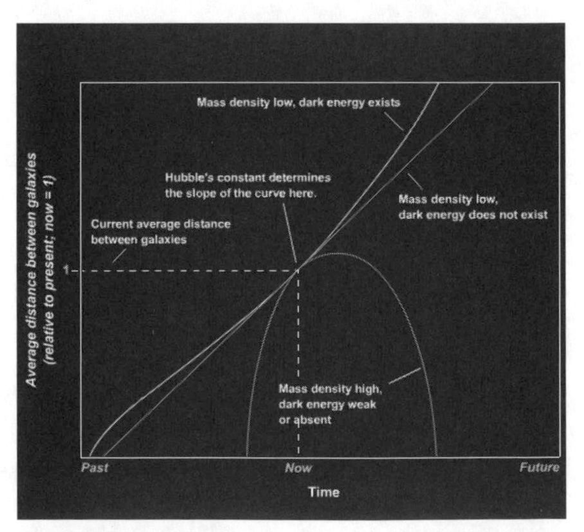

Figure 22-3 The age of the universe depends on mass density and the amount of dark energy present.

LAB ACTIVITY Fate of the Universe

The purpose of this activity is to determine the possible fates of the universe based on differing estimates of both its mass density and the strength of dark energy.

Use the tool from Lesson 3 (shown in Figure 22-4) to adjust the mass density and strength of dark energy to the values in the table. Then, for each case, record the age of the universe and its fate (open, closed, or flat), using the following definitions:

Open: The universe will expand at an increasing rate

Closed: The universe will expand at a decreasing rate

Flat: The universe will expand at a constant rate.

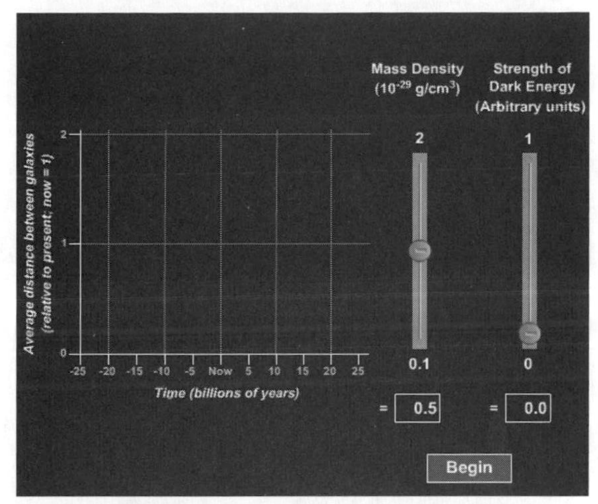

Figure 22-4 This tool from Lesson 3 shows how adjusting the mass density and the strength of dark energy affects the age and ultimate fate of the universe.

Mass Density	Strength of Dark Energy	Age of Universe	Fate
0.3	0		
0.3	1		
0.3	0.7	14 billion years	open
2	0.7		
1	0.7		

For the next-to-last row, try to find the mass density and strength of dark energy that give the oldest possible age for the universe.

For the last row, try to find the mass density and strength of dark energy that will give a closed universe.

Based on your results, which fate do you think is most likely for our universe? Explain your choice.

Voyager: SkyGazer

CD-ROM ACTIVITIES

1
Introducing *SkyGazer*

INTRODUCTION

Voyager: SkyGazer is an interactive desktop planetarium program. It is specifically designed to teach students about astronomy in an interactive environment.

You can use *Voyager: SkyGazer* to see a picture of the sky as it would look from any location at any time. You can also use *Voyager: SkyGazer* to simulate many other astronomical events. This first activity is an introduction and guide to the program. The subsequent activities are designed to help you understand various astronomical concepts through the simulation of events.

PART 1: OPENING *VOYAGER: SKYGAZER*

Open the SkyGazer folder by double-clicking on it. See Figure 1-1 below.

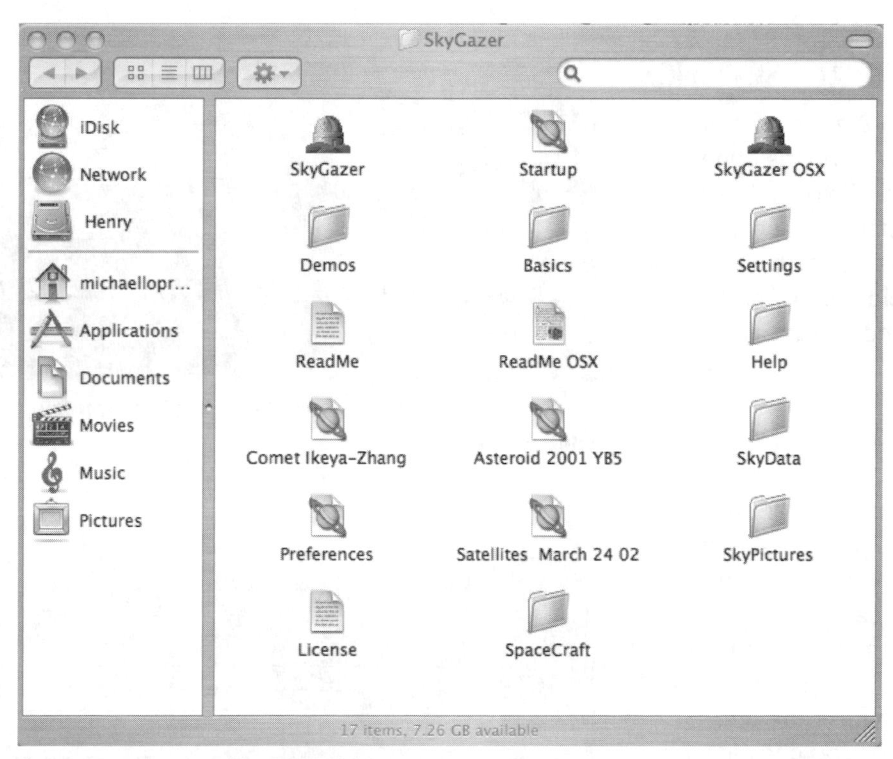

Figure 1-1 The open **SkyGazer** folder

Note: The screen shots in *Voyager: SkyGazer* Activities were taken using the Apple Macintosh version of the software. Make sure to click on the *SkyGazer* OSX icon if you are using system 10 (click on *SkyGazer* icon if you still use system 9). If you're running the Windows operating system, the appearance of the interface will differ slightly.

To open *Voyager: SkyGazer*, double-click on the program icon, labeled **SkyGazer**. When the program opens, you should see a Sky Chart similar to the one shown in Figure 1-2. *Note:* When opening, *SkyGazer* will default to the date and time on your computer's system clock, so your Sky Chart will not look exactly like Figure 1-2. Depending on when you do, it may show a daytime or nighttime sky.

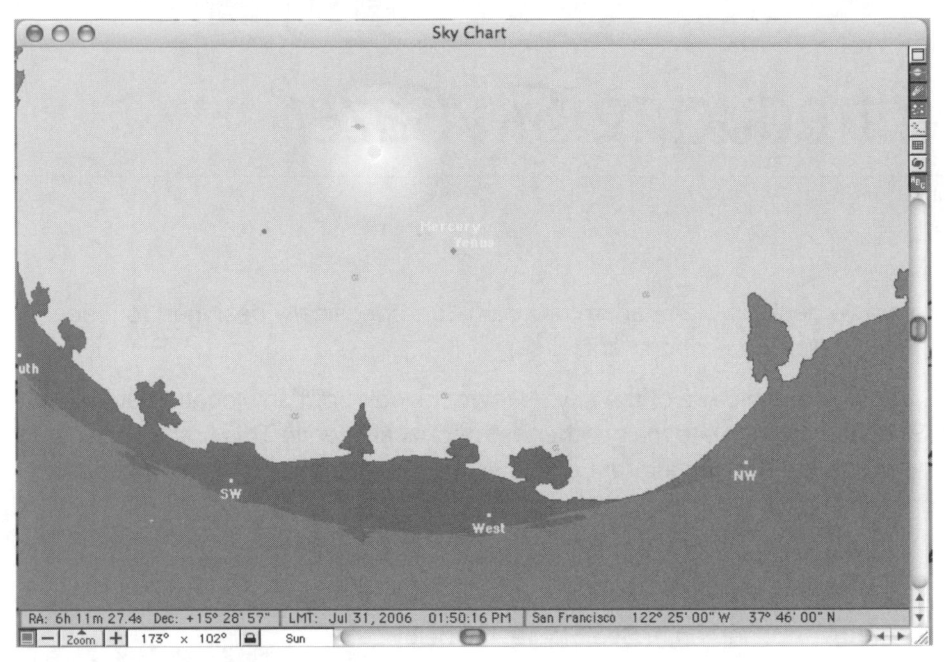

Figure 1-2 The Sky Chart

PART 2: *VOYAGER: SKYGAZER* SETTINGS

The Sky Chart is *Voyager: SkyGazer's* main display. You can adjust the settings by using the panels found under the **Control** menu. To open the **Time Panel**, click on the **Control** menu, hold the mouse button down, and select **Time Panel**. Do the same for the **Location Panel**, **Display Panel**, and **Planet Panel**, respectively. These panels are shown in Figures 1-3 and 1-4.

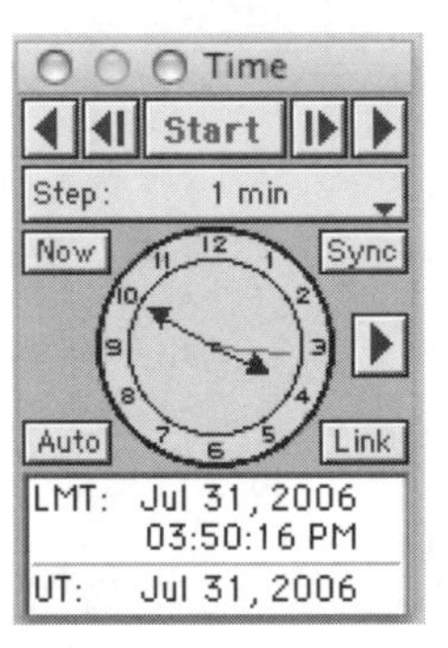

Figure 1-3A The **Time Panel**

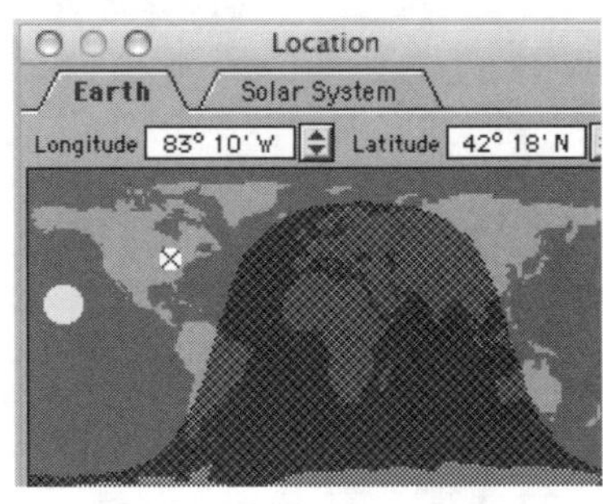

Figure 1-3B The **Location Panel**

SETTING YOUR LOCATION

One of *Voyager: SkyGazer's* most important features is that it displays which objects are visible in the sky at any given time from any given location. To set your location, select **Set Location** under the **Control** menu. When the **Set Location** dialog box (shown in Figure 1–5A) appears, you can specify the **Latitude** and **Longitude** of your location, or you can click on the **List Cities** button. When a list of cities, a portion of which is shown in Figure 1–5B, appears, simply choose a major city near your location.

Important! You cannot change location by typing the name of a city in the **Name** field of the

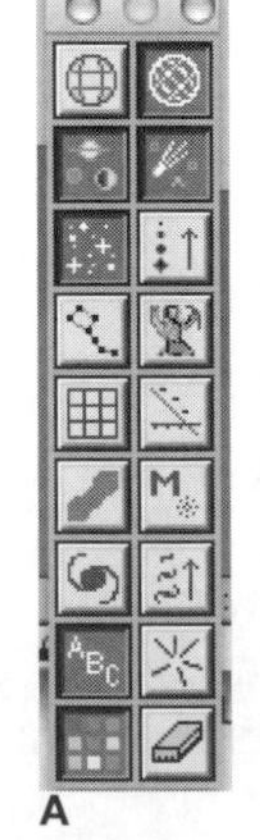

Planet Panel

Planets	Name	Sym	Orbit	Path	Trail
Sun	X	X			
Mercury	X	X			
Venus	X	X			
Earth		X			
Mars	X	X			
Jupiter	X	X			
Saturn	X	X			
Uranus	X	X			
Neptune	X	X			
Pluto	X	X			
Moon	X	X			
Shadow		X			

A B

Figure 1-4 (A) The **Display Panel**, **(B)** The **Location Panel**

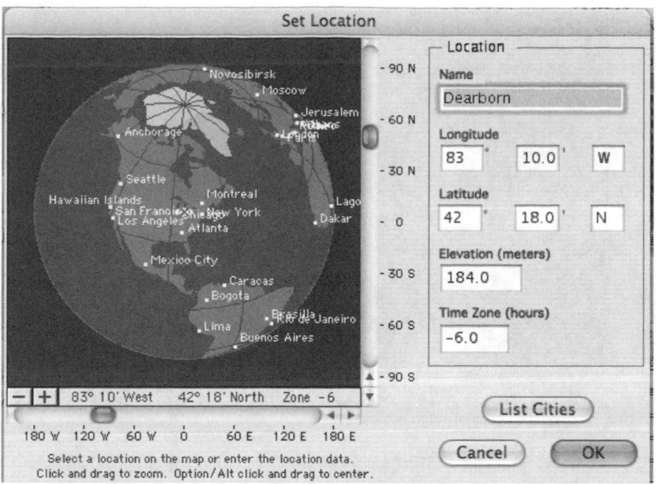

Figure 1-5A The **Set Location** dialog box

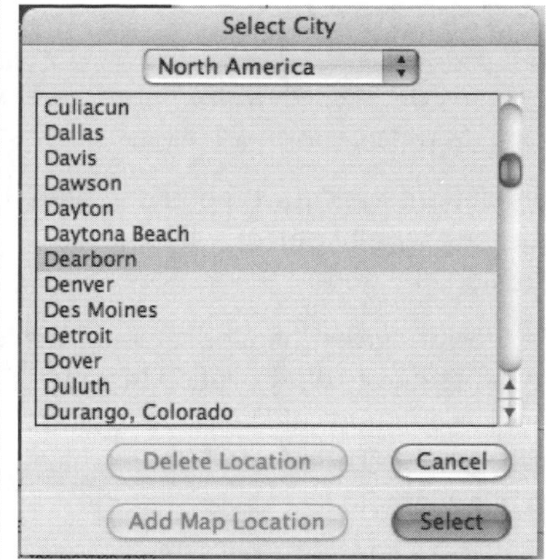

Figure 1-5B The **Select City** dialog box

Set Location dialog box. You must use **List Cities** and then select the city you want.

You can also use the **Location Panel**, found under the **Control** menu (shown in Figure 1-6). To change your location, click on the **location marker** ⊕, hold down the mouse button, and drag the ⊕ to a new location.

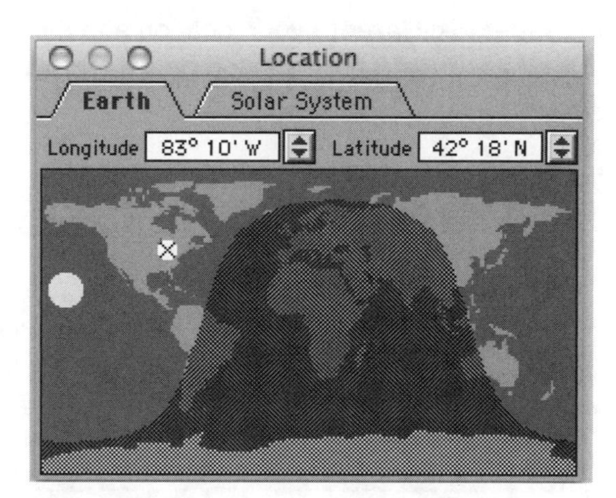

Figure 1-6 Changing your location with the **Location Panel**

SETTING YOUR TIME AND DATE

Now set the time and date of your observation by selecting **Set Time** under the **Control** menu. When the dialog box in Figure 1-7 appears, you can enter any time or date that you want.

The **Time Panel**, also found under the **Control** menu, can be used to move the time forward or backward. The button marked **Step** in Figure 1-8 is used to select the time step. With a larger time step, time on the Sky Chart will pass more quickly than with a smaller time step. When you click on this button and hold down your mouse button, a palette will open, with a wide variety of time steps from which to choose. The arrows framing the **Start** button are used to advance time. The inner arrows move time one step at a time. The outer arrows will move time automatically. The right arrows move time forward, the left arrows move time backward. You can also change the time by dragging the hands of the clock.

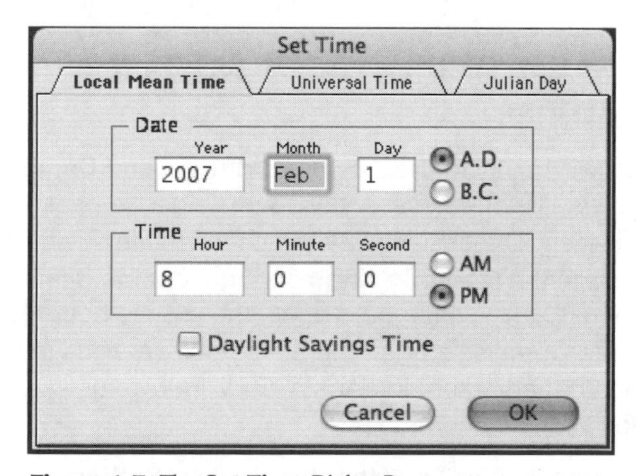

Figure 1-7 The **Set Time** Dialog Box

Figure 1-8 The Time Panel

SKY CHART SETTINGS

Figure 1-9 shows the Sky Chart for February 1, 2007, at 8 P.M. from Dearborn, Michigan.

Chart Labels under the **Chart** menu can be used to label any type of object. Planets and stars are labeled in Figure 1-9.

Magnitude Limits, under the **Chart** menu, will simulate the amount of light pollution in your sky. Dearborn is a suburb of Detroit, Michigan, so **Large City** under **Magnitude Limits** was selected in our example.

Click on any object in the Sky Chart. This will bring up its **Data Panel**, which provides information about the object and is shown in Figure 1-10.

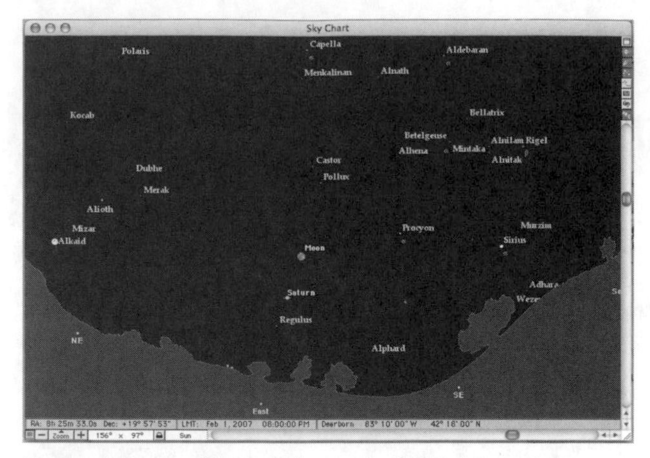

Figure 1-9 Sky Chart for February 1, 2007 at 8 P.M. from Dearborn, Michigan

Set *Voyager: SkyGazer* to your location and the current date. See what is in the sky tonight at 10:00 P.M. Record the requested information on the RESULTS sheet at the end of this activity.

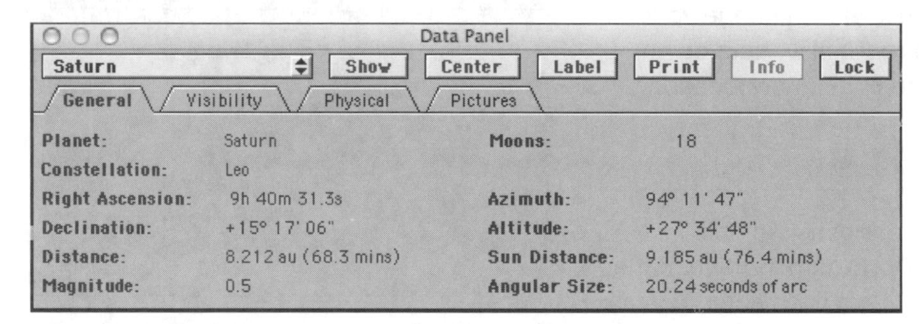

Figure 1-10 Click once on an object to bring up its **Data Panel**

PART 3: USING *VOYAGER: SKYGAZER* FEATURES

You will be using the **Basics**, **Settings**, and **Demos** folders, all found in the **SkyGazer** folder, to begin most activities. The easiest way to access these settings is through the **Open Settings** command under the **File** menu. Click on **File**, hold down the mouse button, and select **Open Settings**. The dialog box seen in Figure 1-11 should then come up. (Again, for Windows users this will look slightly different.)

You can then select either the **Basics**, **Settings**, or **Demos** folder from the left side of the dialog box by clicking on it. The folder will open and you can select the setting you wish to use from the right side of the dialog box. The dialog box will then appear as in Figure 1-12.

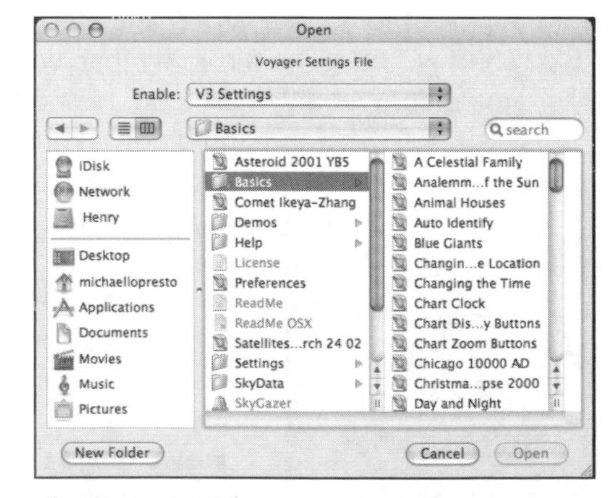

Figure 1-11 Dialog Box under the **Open Settings** command

Many of the settings in the **Basics** folder are designed to help you learn to use *Voyager: SkyGazer*. To conclude this introduction, select in turn each of the settings on the RESULTS sheet. Follow their instructions and record on the RESULTS sheet at the end of this activity what you saw or what you learned about each feature of *Voyager: SkyGazer*.

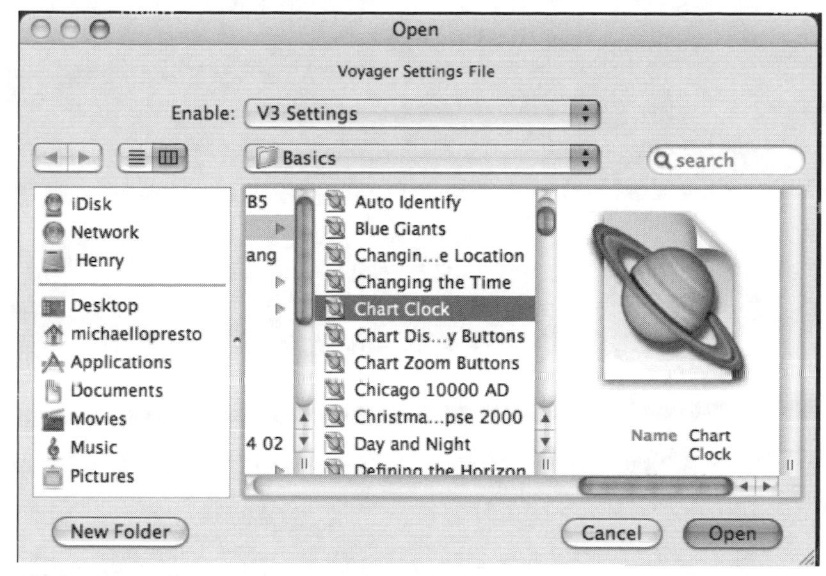

Figure 1-12 Settings found in the **Basics** folder

RESULTS SHEET 1 Introducing *SkyGazer*

NAME ______________________ **DATE** ______________ **SECTION** ______________________

PART 2: *VOYAGER: SKYGAZER* SETTINGS

Your Location: ______________________________

Current Date: ______________________________

Which **Magnitude Limits** setting did you use under the **Chart** menu?

List all planets, if any, that are visible in your sky tonight.

Click on one of the planets to bring up its **Data Panel**.

Under the **General** tab, find and record the following information:

Constellation the planet is in: ____________________

Its **Angular Size**: ______________________________

Under the **Visibility** tab, find and record the following information:

Rise Time: ____________________

Set Time: ____________________

Under the **Physical** tab, find and record the following information:

Mass: ____________________

Diameter: ____________________

Moons: ____________________

Under the **Pictures** tab, find and look at the pictures of the planet.

List four bright stars that are [illegible]. Click on each of them to bring up a **Data Panel** and record their information in the table below.

Star Name	In Constellation	Midnight Transit	Magnitude

PART 3: USING *VOYAGER: SKYGAZER* FEATURES

File | Open Settings | Basics

Select each setting in the **Basics** folder and follow the directions given. Describe what you saw or what you learned about each function of *Voyager: SkyGazer*.

Grid Lines

Planet Panel

Chart Clock

Shortcuts

Chart Display Buttons

Sky Labels

Chart Zoom Buttons

Defining the Horizon

Day and Night

Sky Scrolling

Sky Zoom

Dragging the Sky

CONCLUSION

In the space below, write a conclusion for this activity. Briefly explain what you did and what you learned from it.

CHECK YOUR UNDERSTANDING 1: INTRODUCING *SKYGAZER*

MULTIPLE-CHOICE QUESTIONS

1. Your time and location are set from under the ________________ menu.
   a. Control
   b. Chart
   c. View
   d. Display
2. A single-click on an object will
   a. make it disappear.
   b. center on it.
   c. bring up its Data Panel.
   d. [both b and c].
3. A double-click on an object will
   a. make it disappear.
   b. center on it.
   c. bring up its Data Panel.
   d. [both b and c].
4. Which panel can be used to turn the constellation figures on and off?
   a. Display
   b. Time
   c. Planet
   d. Location
5. Which panel can be used to center and lock on an individual planet?
   a. Display
   b. Time
   c. Planet
   d. Location
6. You can set your location with
   a. the Location Panel.
   b. with the Set Location . . . command under the Control menu.
   c. by clicking on the Earth in the Set Location dialog box.
   d. [all of the above].

7. Which can be done with the Time Panel, but not with the Set Time . . . command under the Control menu?
   a. set the time of day
   b. set the date
   c. set the time step interval
   d. [all of the above]
8. Which is NOT a choice for Horizon Visibility?
   a. transparent
   b. translucent
   c. opaque
   d. invisible
9. Which is NOT a method for zooming in or out?
   a. the + and − buttons along the bottom scroll bar
   b. selecting an angle under the Zoom button
   c. dragging the cursor over a section of the Sky Chart
   d. [all of the above CAN be used for zooming]
10. In which folder will you find settings ready for you to use?
   a. Basics
   b. Demos
   c. Settings
   d. [all of the above]

OPEN-ENDED ACTIVITY

Using the skills you learned in this activity, find and report the name and type of three different objects (one each at sunset, midnight, and sunrise) that can currently be seen high overhead from your location.

VOYAGER: SKYGAZER CD-ROM ACTIVITIES

2

Motions of the Stars

INTRODUCTION

In this activity, you will learn about the motions of the stars that can be observed from your location and other locations on Earth. Before beginning this activity, read the section on Altazimuth Coordinates found on pages 3–5 of *Appendix A: Basic Concepts*. If you need detailed instructions on how to change the Sky Chart settings such as time, date, and location, refer to *Voyager: SkyGazer Activity 1, Introducing SkyGazer.*

PART 1: OBSERVING THE MOTIONS OF THE STARS FROM HOME

File | Open Settings | Basics | Three Cities

1. Close all the windows except the one titled **Sydney** and the **Time Panel**.
2. Use the **Time Panel** to select a time step of **1 minute**.
3. Select **Set Location** . . . under the **Control** menu and set the location to a city near your location. Also, open the **Location Panel*** under the **Control** menu to see your location.
4. Select **Set Time** . . . under the **Control** menu and change the time to the current date at any time of day.
5. Under the **Chart** menu, deselect **Natural Sky**. This will take away the display's changes between day and night, which is better for this activity. Also select **Chart Labels** under the **Chart** menu and deselect **Star Names**.
6. Zoom out by clicking on the **Zoom** button at the bottom left of your Sky Chart until your chart looks similar to Figure 2-1. Ignore the fact that your chart is still labeled "Sydney."
7. Now use the **Time Panel** to advance time in 1-minute time steps. To manually advance time, use the inner arrows on either side of the start button. Watch the motion of the sky.

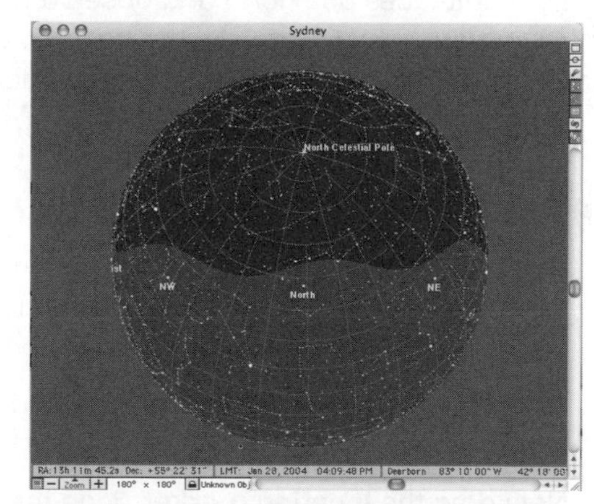

Figure 2-1 Your Sky Chart after steps 1–6 in **Part 1**

8. Watch the stars as they appear to move, observing first due east, then due west, due south, and finally due north. Use the up and down and back and forth scroll bars on your Sky Chart or your keyboard's up, down, left, and right arrow keys to look in the different directions. On the RESULTS

*A shortcut for bringing up or turning off the Time, Location, Display, and Planet Panels is to press the 'T,' 'L,' 'D,' and 'P' keys, respectively.

sheet, draw an arrow that represents the direction of the stars' motion when viewed in each of these four directions. It may help to focus on the motion of a single star.

9. Note the point above the horizon that the sky seems to be moving around and the star very close to that point. Click on that star to bring up its **Data Panel** and record on the RESULTS sheet the name of that star and its *altitude*—how high above the horizon it is.

10. Use the **Location Panel** to record your latitude.

PART 2: USING THE SKY TO DETERMINE YOUR LOCATION

The spot marked ⊗ on the **Location Panel** is your **location marker**. By clicking on this point and holding the mouse button down, you can drag the **location marker** and set your position to anywhere on the Earth.

1. Click and drag your location north and south, east and west, and observe what happens. Pay special attention to the north celestial pole in the sky when you move north or south.

2. Change your location to the North Pole, 90° north latitude. Once you drag your location marker to the top of the **Location Panel** display, you can increase your latitude on the **Location Panel** up to 90° by clicking on the arrows next to the **Latitude** box. Now click on the pole star you have identified to bring up its **Data Panel** again and record its altitude when seen from the North Pole on your RESULTS sheet.

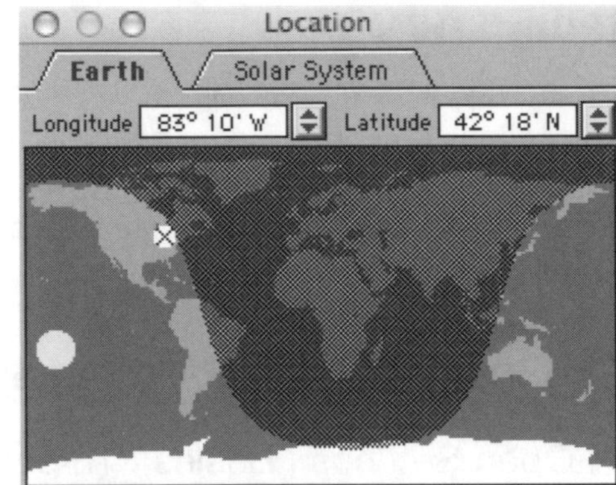

Figure 2-2 The Location Panel

3. Advance time and watch the motion of the stars. On the RESULTS sheet, draw an arrow that represents the direction of motion of the stars when viewed from the North Pole.

4. Move your latitude south by slowly dragging the **location marker** ⊗ on the **Location Panel** downward. With the Sky Chart's scroll bars or your computer's arrow keys, adjust your Sky Chart so that it faces due north and observe what happens to the north celestial pole. Stop moving south when you reach 0° latitude, the equator. Click on the pole star and record its altitude on the RESULTS sheet. Also, draw an arrow that represents the direction of the motion of the stars when viewed from the equator.

PART 3: WHICH STARS CAN YOU SEE?

1. Use the **Location Panel** to return to your original position and once again advance time to observe the motion of the stars. This time take notice of the stars as they rise and set. Answer the questions on the RESULTS sheet.

2. Do the same thing at the North Pole and the equator and answer the questions on the RESULTS sheet.

3. Now drag the **location marker** ⊗ on the **Location Panel** to a location in the southern hemisphere that is comparable in southern latitude to the northern latitude of your home location. Observe the motion of the stars there and answer the questions on the RESULTS sheet.

4. Predict what you will observe from the South Pole and fill in the RESULTS sheet accordingly. Then, go to the South Pole and see if you were right.

RESULTS SHEET 2 Motions of the Stars

NAME ____________________ **DATE** ____________ **SECTION** ____________

PART 1: OBSERVING THE MOTIONS OF THE STARS FROM HOME

Draw an arrow in each figure to represent the direction of the stars' motion that you observed. Do not be afraid to zoom in if necessary.

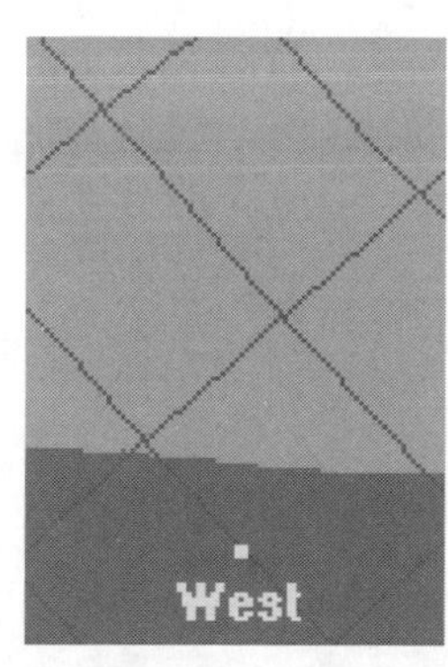

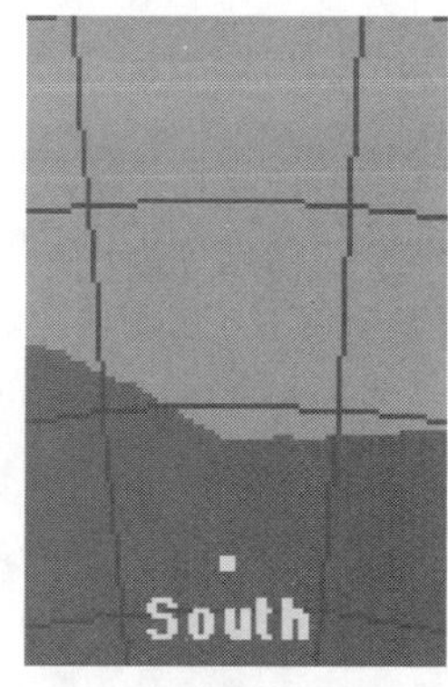

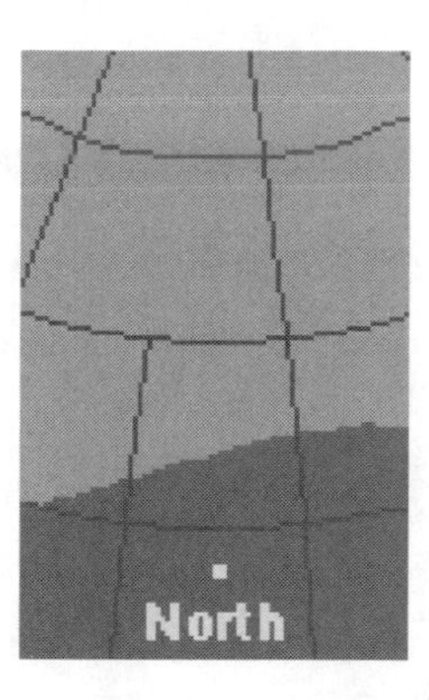

Figure 2-3 Draw an arrow in each panel to represent the direction of the stars' motion

Name of the point around which all the stars seem to move:

Name of a star near that point:

Altitude of that pole star (from its **Data Panel**):

Your location:

Your latitude (from **Location Panel**):

PART 2: USING THE SKY TO DETERMINE YOUR LOCATION

Altitude of the pole star as seen from the North Pole:

Latitude of the North Pole:

Altitude of the pole star as seen from the equator:

Latitude of the equator:

Draw an arrow in each panel of Figure 2-4 to represent the direction of the stars' motion that you observed. Do not be afraid to zoom in if necessary.

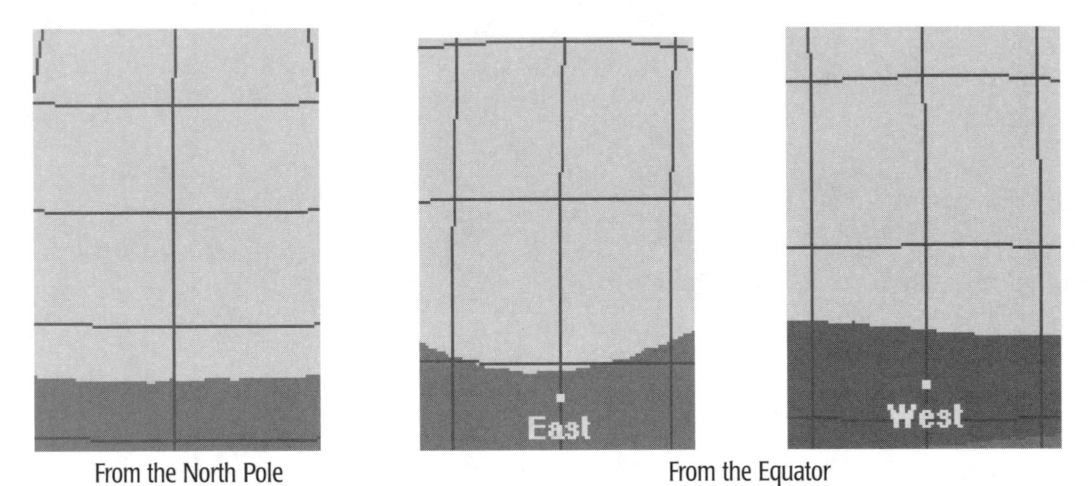

From the North Pole — From the Equator

Figure 2-4 Draw an arrow in each panel to represent the direction of the stars' motion that you observed at each location.

Complete the following table:

Location	Latitude	Altitude of Pole Star
North Pole		
Your Location:		
Equator		

What is the relationship between the observer's latitude and the pole star's altitude?

PART 3: WHICH STARS CAN YOU SEE?

Do stars rise and set in your home location?

Are there stars that never set in your home location? (These are known as *circumpolar stars.*)

Are there stars that never rise in your home location?

Transfer your answers to the last three questions to the Your Location row of the table below. Fill in the name of your home location in the Location column. Based on your observations, fill out each space in the table for the North Pole and equator with either "Yes" or "No."

Location	Rise and Set?	Never Set?	Never Rise?
North Pole			
Your Location:			
Equator			
Southern Location:			
South Pole			

Fill in the name of the southern hemisphere location you chose in the Location column and again, based on your observations, fill out each space in the table with "Yes" or "No" for that location.

What do you notice about the stars that never set and never rise at the southern hemisphere location compared to your home location?

Predict how the stars will move when observed from the South Pole. Fill in the table with your predictions.

Go to the South Pole. Was your prediction correct?

After observing the motions of the stars from the South Pole, name one similarity and one difference compared to the motion you observed at the North Pole.

CONCLUSION

In the space below, write a conclusion for this activity. Briefly explain what you did and what you learned from it.

CHECK YOUR UNDERSTANDING 2: MOTIONS OF THE STARS

MULTIPLE-CHOICE QUESTIONS

1. The altitude of the pole star
   a. is always 90°.
   b. is the same as the observer's latitude.
   c. is the same for all northern hemisphere observers.
   d. is exactly the same as the altitude at the north celestial pole.
2. Stars near the north celestial pole appear to move
   a. in lines parallel to the horizon.
   b. in lines perpendicular to the horizon.
   c. in concentric circles around the north celestial pole.
   d. in east-to-west arcs centered below the horizon.
3. When observed above due south on the horizon, stars appear to move
   a. in lines parallel to the horizon.
   b. in lines perpendicular to the horizon.
   c. in concentric circles around the north celestial pole.
   d. in east-to-west arcs centered below the horizon.
4. When observed from the equator, stars appear to move
   a. in lines parallel to the horizon.
   b. in lines perpendicular to the horizon.
   c. in concentric circles around the north celestial pole.
   d. in east-to-west arcs centered below the horizon.
5. When observed from the North Pole, stars appear to move
   a. in lines parallel to the horizon.
   b. in lines perpendicular to the horizon.
   c. in concentric circles around the north celestial pole.
   d. in east-to-west arcs centered below the horizon.
6. If stars appear to move toward the horizon, you are looking
   a. north.
   b. south.
   c. east.
   d. west.

7. If stars appear to move downward at an angle less than 90° toward the horizon, you could be at
   a. the North Pole.
   b. the equator.
   c. a middle-north latitude.
   d. [none of the above].
8. When observed at a southern mid-latitude, stars appear to
   a. rise in the west.
   b. move in circles around a point above the northern horizon.
   c. reach their highest altitude above the northern horizon
   d. [all of the above].
9. About what percentage of the stars visible from Earth can be seen from the equator?
   a. 25%
   b. 50%
   c. 75%
   d. 100%
10. About what percentage of the stars visible from Earth can be seen from both the North and the South Poles?
   a. 0%
   b. 50%
   c. 100%
   d. [it depends on the time of year]

OPEN-ENDED ACTIVITY

Using the skills you learned in this activity, identify a constellation that is circumpolar from your location. Also identify a rising and setting constellation that will currently be visible from your location at some time.

VOYAGER: SKYGAZER CD-ROM ACTIVITIES

3
The Celestial Sphere

INTRODUCTION

In this activity, you will learn about the celestial sphere and see that it is a useful model for predicting which parts of the sky you will see from various locations on the Earth. Before beginning this activity, read the section on the celestial sphere found on pages 1–2 of *Appendix A: Basic Concepts*.

PART 1: COORDINATES ON THE CELESTIAL SPHERE

File | Open Settings | Basics | Three Cities

1. Close all the windows except the one titled **Sydney**. Under **Sky View** under the **Chart** menu, select **Star Atlas Coordinates**.
2. Set your location to a city near you. Open the **Location Panel** to see your location. Remember that you also can use the **location marker** ⊕ on the **Location Panel** to set your location.
3. Select **Reference Markers** under the **Display** menu and deselect all markers except **Celestial Equator** and **Celestial Poles**. A check mark indicates that something is selected. Zoom out; your Sky Chart should look similar to the one in Figure 3-1.

You can use your computer's scroll bars or arrow keys to change your view of the celestial sphere.

The celestial sphere is an imaginary Earth-centered sphere to which all celestial objects appear to be attached. All points on the celestial sphere correspond to the points on the Earth that they are directly above. For instance, the *north celestial pole* is the point on the celestial sphere directly above the Earth's North (geographic) Pole. The *celestial equator* is the line on the celestial sphere directly above the Earth's equator at all points.

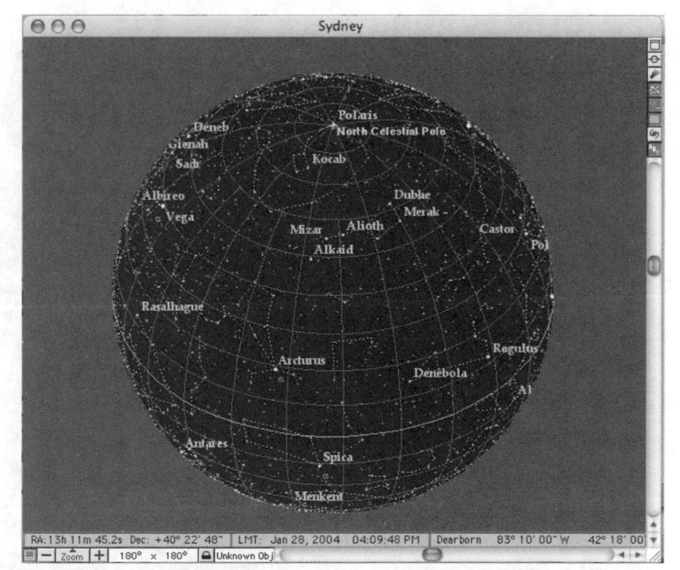

Figure 3-1 Your Sky Chart after Steps 1–3 in Part 1

The up and down scroll bar or arrow keys adjust the declination you are viewing. *Declination* is a measurement of how many degrees a point is north (+) or south (−) of the celestial equator. It is very similar to latitude on the Earth.

4. Use the up and down scroll bar or arrow keys to change the declination at the center of your Sky Chart. Record the declinations of the objects listed on the RESULTS sheet.

   The right and left scroll bar or arrow keys adjust the right ascension you are viewing. *Right ascension* is a measurement of east and west on the celestial sphere. It is similar to longitude on Earth, but it is measured in hours rather than in degrees. This is because once a day, the celestial sphere appears to move around the Earth, so there are 24 hours of right ascension.

5. Use the right and left scroll bar or arrow keys to change the right ascension that is at the center of your Sky Chart. Set the declination of the center of your chart to 0°, the celestial equator, and record the right ascensions of the objects listed on the RESULTS sheet.

   Together, declination and right ascension are known as *equatorial coordinates*. The grid shown on your Sky Chart in Figure 3-1 is equatorial. The lines of equal declination are circles around the celestial poles parallel to the celestial equator. The lines of equal right ascension run between the celestial poles and are farthest apart at the celestial equator.

   See *Appendix A: Basic Concepts* for a review of equatorial coordinates.

PART 2: THE COORDINATES OF STARS

1. Find the star *Mirphak* in the constellation Perseus.
2. Double-click on *Mirphak* to center on it and bring up its **Data Panel**. Record its declination and right ascension on the RESULTS sheet.
3. Use the left and right scroll bar or arrow keys to find another labeled star that is nearly on the same line of declination as *Mirphak* and record its coordinates.
4. Find the star *Almaak* in Andromeda. Double-click on *Almaak* to center on it and bring up its **Data Panel**. Record its right ascension and declination on the RESULTS sheet.
5. Now use the up and down scroll bar or arrow keys to find a labeled star just south of *Almaak* near the same line of right ascension and record its coordinates.

PART 3: GLOBAL AND LOCAL VIEWS

1. Switch to **Local Horizon Coordinates** from the **Chart** menu under **Sky View**.
2. Also from the Chart menu, make sure Natural Sky is deselected.
3. Under the **Display** menu enable these **Reference Markers**: **Horizon Line**; **Meridian Line**; **Zenith and Nadir**; and **Cardinal Points**. Make sure that the **Celestial Equator** and the **Celestial Poles** are selected as well.

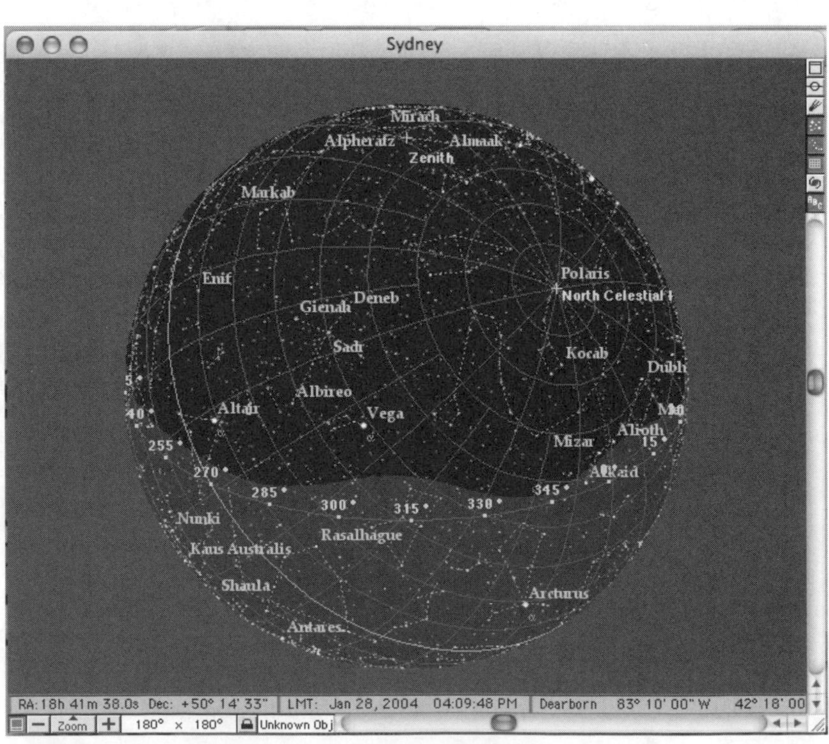

Figure 3-2 Your Sky Chart after Steps 1–3 in Part 3

Your Sky Chart should look similar to Figure 3-2 (although the position of the stars will vary with location).

You have changed your point of view from a *global* one, the celestial sphere, to the *local* view at your location. Equatorial coordinates are global and they can still be used at your location, but it is convenient to have a local coordinate system as well. These are known as *altazimuth coordinates*. They are also often called horizon coordinates.

Now that you have changed your **Sky View** from **Star Atlas Coordinates** to **Local Horizon Coordinates**, your scroll bars and arrow keys change the altazimuth coordinates of your view. The up and down scroll bar and arrow keys change the altitude of the center of your Sky Chart and the back and forth scroll bar and arrow keys change the azimuth of the center of your chart. *Altitude* is a measurement of how many degrees above the horizon an object is. *Azimuth* is a measurement of how many degrees along the horizon away from due north the point on the horizon directly below an object is. Due north is azimuth 0°; due east, 90°; due south, 180°; and due west, 270°.

See *Appendix A: Basic Concepts* for a review of altazimuth coordinates.

The Sky Chart shown on the previous page is the sky as seen from Detroit, Michigan. The corresponding **Location Panel**, shown in Figure 3-3, is also set for Detroit.

Detroit is an example of a middle-north latitude. Most people (and probably you) live in a middle-north latitude.

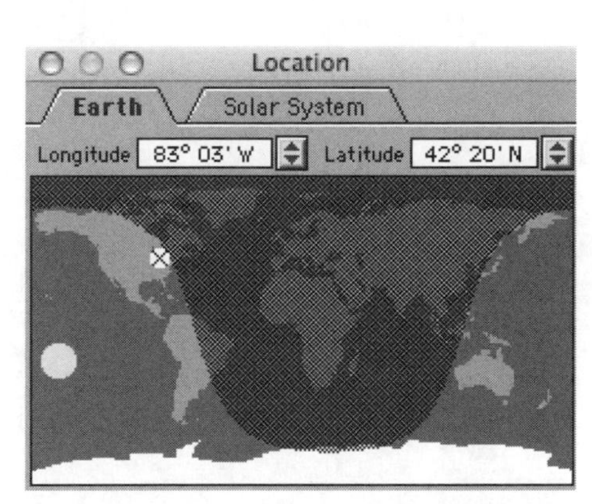

Figure 3-3 The Location Panel set for Detroit, Michigan

4. Use your cursor to click on your **location marker** ⊕ and drag it back and forth east and west *only*. Notice what changes you see. Now drag the **location marker** ⊕ north and south and note the changes. Record the changes you observe on the RESULTS sheet.
5. Look carefully at the Sky Chart as you use the **Location Panel** to change your location. Try to figure out which reference markers *do* and *do not* move as you change location.

PART 4: VIEWS FROM AROUND THE WORLD

1. Place your **location marker** ⊕ on the **Location Panel** at a middle-north latitude (your approximate location should do fine). What relationship do you expect between your latitude and the altitude of the north celestial pole? Record your prediction on the RESULTS sheet. Your latitude can be found on the **Location Panel** and the approximate altitude of the north celestial pole can be found by clicking on Polaris and bringing up its **Data Panel**.
2. Now, by clicking on the latitude arrows on the **Location Panel**, slowly increase your latitude. This means you are moving north. Make sure you can see the north celestial pole and the zenith. If you cannot, use the scroll bars or arrow keys to move them into view. Observe what happens to the celestial sphere as you move north. Record any changes on the RESULTS sheet.
3. Now slowly move south from the North Pole. Continue past 0° latitude, the equator, and eventually come to the South Pole. Stop and record on the RESULTS sheet your observations from the locations.

RESULTS SHEET 3 The Celestial Sphere

NAME ______________________ **DATE** ______________ **SECTION** ______________________

PART 1: COORDINATES ON THE CELESTIAL SPHERE

Zoom in and click on the following stars. Record their declination (Dec.):

Star near the north celestial pole: ____________________

This star's name: ____________________

Capella: ____________________

Alnilam (near the celestial equator, in Orion's belt): ____________________

Canopus: ____________________

Now record the right ascension (RA) of the following stars:

Arcturus: ____________________

Regulus: ____________________

Aldebaran: ____________________

Markab: ____________________

Vega: ____________________

Between which two of the above stars would the line of 0 hrs right ascension be?

PART 2: THE COORDINATES OF STARS

In the table below, enter the information for the star with a similar declination to *Mirphak* and the star with a similar right ascension to *Almaak*. Zoom in on the star to make it easier to click on it to bring up the **Data Panel**.

Star	Declination	Right Ascension
Mirphak		
Star with *Mirphak's* declination:		
Almaak		
Star with *Almaak's* right ascension:		

PART 3: GLOBAL AND LOCAL VIEWS

Describe the changes you see in the stars and the reference markers when you:

change location from east to west.

change location from north to south.

Circle the names of the reference markers that move when you change location. *Do not* circle those that remain stationary.

Lines	***Points***
Horizon	**Zenith**
Meridian	**Cardinal Points**
Celestial Equator	**Celestial Poles**

The reference markers that *do* move as you change location are part of the **EQUATORIAL** | **ALTAZIMUTH** *(circle one)* coordinate system. Those that *do not* move as you change location are part of the **EQUATORIAL** | **ALTAZIMUTH** *(circle one)* coordinate system. Therefore, equatorial coordinates are a **GLOBAL** | **LOCAL** *(circle one)* system and altazimuth coordinates are **GLOBAL** | **LOCAL** *(circle one)*.

PART 4: VIEWS FROM AROUND THE WORLD

Your Latitude: ____________________

Predicted Altitude of Polaris: _____________

Actual Altitude of Polaris: _____________

Explain the reasoning behind your prediction:

Answer the following questions by changing your latitude on the **Location Panel** and observing the changes that occur on your Sky Chart.

As you move north, describe what happens to:

the north celestial pole.

the celestial equator.

Once you get to the North Pole, 90°N latitude, the north celestial pole is seen at the ____________.
The celestial equator is seen at the ____________.

As you move south, describe what happens to:

the north celestial pole.

the celestial equator.

Once you get to the equator, 0° latitude, the north celestial pole is seen at the ____________.
The celestial equator is seen going through the ____________.

In the southern hemisphere the altitude of the ____________ will be equal to your southern latitude. So, as you move south, the altitude of this point **INCREASES** | **DECREASES** *(circle one)*.

Once you get to the South Pole, 90°S latitude, the ____________ is seen at the zenith.
The ____________ is seen on the horizon.

CONCLUSION

In the space below, write a conclusion for this activity. Briefly explain what you did and what you learned from it.

CHECK YOUR UNDERSTANDING 3: THE CELESTIAL SPHERE

MULTIPLE-CHOICE QUESTIONS

1. As you change locations, which of a star's coordinates will change?
   a. right ascension and declination
   b. altitude and azimuth
   c. latitude and longitude
   d. [none of the above will change]
2. If you want to tell a friend who lives at a different latitude than you where to look for an object in the sky, you should give them the object's
   a. right ascension and declination.
   b. altitude and azimuth.
   c. latitude and longitude.
   d. [any of the above will be the same].
3. Which of the following reference markers are NOT local?
   a. zenith
   b. meridian
   c. cardinal points
   d. north celestial pole
4. Which of the following reference markers are NOT global?
   a. horizon
   b. celestial equator
   c. north celestial pole
   d. [all three ARE global]
5. Which is most similar to latitude?
   a. right ascension
   b. declination
   c. altitude
   d. azimuth
6. Which is most similar to right ascension?
   a. latitude
   b. longitude
   c. altitude
   d. declination

7. Which is NOT measured in degrees?
   a. right ascension
   b. declination
   c. altitude
   d. latitude
8. The ____________ of the pole star is equal to the ____________ of the observer.
   a. latitude, altitude
   b. declination, latitude
   c. altitude, latitude
   d. latitude, declination
9. The declination of the celestial equator is
   a. +90°.
   b. −90°.
   c. 0°.
   d. [none of the above].
10. The declination of the north celestial pole is
   a. +90°.
   b. −90°.
   c. 0°.
   d. [none of the above].

OPEN-ENDED ACTIVITY

Use the skills you learned in this activity to determine the lowest declination that is visible from your location and to find an object that is near this declination.

4

Motions of the Sun

INTRODUCTION

In this activity, you will learn about the motions of the Sun that can be observed from your location and other locations on the Earth. You will see how the motions are similar in some ways to those of the stars, but are different in other ways. You also will learn how the Sun's motions define time measurements and how they cause seasonal changes and weather differences on the Earth.

PART 1: THE SOLAR CLOCK

File | Open Settings | Basics | Three Cities

1. Close all the windows except the one titled **Sydney** and the **Time Panel**.
2. Move the location marker on the **Location Panel*** to a place near your location.
3. Set the date to March 21 of this year at noon. You can also set the time of day by using the cursor to drag the hands of the clock on the **Time Panel**.*
4. Use the Planet Panel under the Control menu to deselect all objects except the Sun. You can select/deselect objects on the Planet Panel by clicking on their buttons. When the buttons are selected, they display white letters on a dark gray background; when deselected, the opposite is true.
5. Turn on the Sun by clicking on the **planet** button (the topmost of the buttons above the up and down scroll bar on your Sky Chart; see Figure 4-1).
6. Use the left and right scroll bar or arrow keys to face south and zoom out until your Sky Chart looks similar to the one in Figure 4-1.
7. Click on the Sun to bring up its **Data Panel**. Under the **Visibility** tab, note the *Transit time*. Reset your clock to this time. This is the time of day that the Sun has its highest altitude in the sky. Record on the RESULTS sheet the transit altitude and the Sun's rise and set times for March 21.

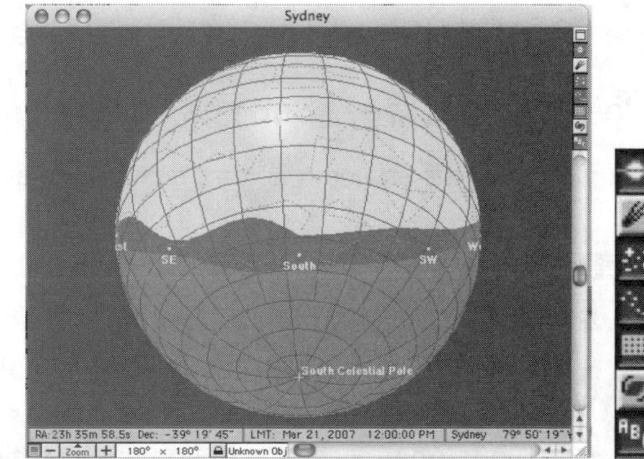

Figure 4-1 Sky Chart facing south at noon on March 21

8. Select **Chart Labels** under the **Chart** menu and deselect **Star Names**. Now manually advance the time with a 1-minute time step and watch the motion of the Sun. Observe the setting point on the western horizon and note whether it is north

*Recall (from Activity 2) the shortcuts for bringing up these panels.

of west, south of west, or due west. *Note:* You will have to use your scroll bars or arrow keys to keep the Sun in your Sky Chart view. Record this information on the RESULTS sheet.

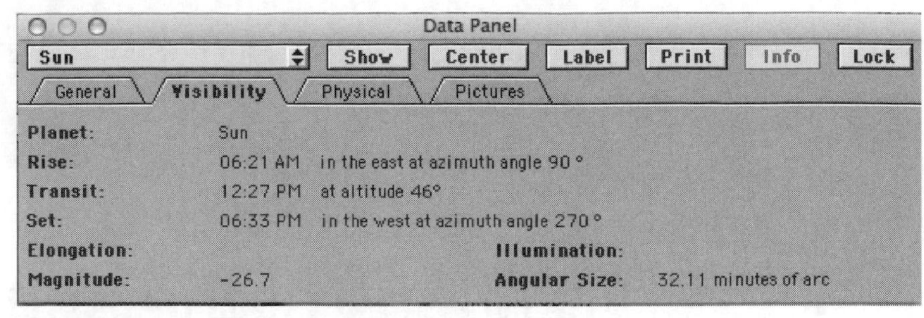

Figure 4-2 The Sun's Data Panel

9. Advance time to the next morning, then record the rising and setting positions of the Sun on the eastern horizon on the RESULTS sheet.

10. Once the Sun returns to transit the next day, one *solar day* has passed. Now change the time step to **1 day** and manually advance time. Note the change(s) you observe in the Sun's midday altitude as you move through the year.

11. Stop advancing time when you get to June 21 and record the same information on the RESULTS sheet that you did for March 21.

12. Continue advancing time, but stop on September 21 and December 21 and record the information on the RESULTS sheet. What change(s) do you notice between June 21 and December 21? Once you have returned to March 21, you have observed the cycle known as a *tropical year*.

PART 2: THE ANALEMMA

Before beginning this section, read about the *analemma* on pages 8–9 of *Appendix A: Basic Concepts*.

1. Double click on the Sun to center it in your Sky Chart, then click on the Sun's box under path. This will display the Sun's path. Advance the time in 1-day steps, this time *not* stopping until a year has passed. The pattern you see, the Sun's path observed once a day, at the same time of day, for a whole year, is called the *analemma*. Draw the pattern you see on the RESULTS sheet.

2. Now open the settings below to see two other analemmas.

File | Open Settings | Basics | Analemma of the Sun

File | Open Settings | Settings | Analemma from Minneapolis

3. Compare the analemmas and answer the questions on the RESULTS sheet.

PART 3: AROUND THE WORLD

1. Reset your Sky Chart in the same way as you did for Part 1, but this time set the date to June 21. Opening the analemma settings above will have wiped out everything you set up before, so you will need to reinstate all the settings.

2. Drag your **location marker** ⊕ on the **Location Panel** north and south and observe what happens to the Sun on the Sky Chart.

Observations for a high northern location:

3. Use the location marker to move your location to a high northern latitude, close to 90°N (the North Pole).

4. Center on the Sun by double-clicking on it or on the word **Sun** on the **Planet Panel**.

5. Advance the time in 10-minute time steps for a whole day and watch the Sun. You will have to use the left and right scroll bars or arrow keys to follow it. Record your observations on the RESULTS sheet.

Observations for an equatorial location:

6. Now use the **Location Panel** to change your latitude to a location near 0° latitude, the equator.
7. Center on the Sun.
8. Once more, advance the time in 10-minute time steps for a whole day and watch the Sun. You will again have to use the left and right scroll bars or arrow keys to follow the Sun. Record your observations on the RESULTS sheet.

Observations for a southern location:

9. Now drag the **location marker** ⊕ on the **Location Panel** to a location in the southern hemisphere that is comparable in latitude to your home location. Make your observations of the Sun and record them on the RESULTS sheet.

Observations from the South Pole:

10. Finally, make and record observations of the Sun from Antarctica, near 90°S, the South Pole.

PART 4: AROUND THE WORLD AGAIN

Repeat steps 3–10 in Part 3 for December 21, then record your observations on the RESULTS sheet. Use what you record to answer the questions on the RESULTS sheet.

RESULTS SHEET 4 Motions of the Sun

NAME ______________________ **DATE** ____________ **SECTION** ______________

PART 1: THE SOLAR CLOCK

Record the Sun's transit altitude, rising and setting times, and rising and setting positions for each date from your location in the table below. Also, compute the number of hours that the Sun is in the sky.

Your location (from **Location Panel**): Longitude: ____________ Latitude: ______________

Date	Transit Altitude	Rise Time	Set Time	Hours in the Sky	Rise Position	Set Position
March 21						
June 21						
Sept 21						
Dec 215						

From inspection of your table, name three ways in which you can observe the passage of a *tropical year*.

PART 2: THE ANALEMMA

Draw the analemmas that you observed:

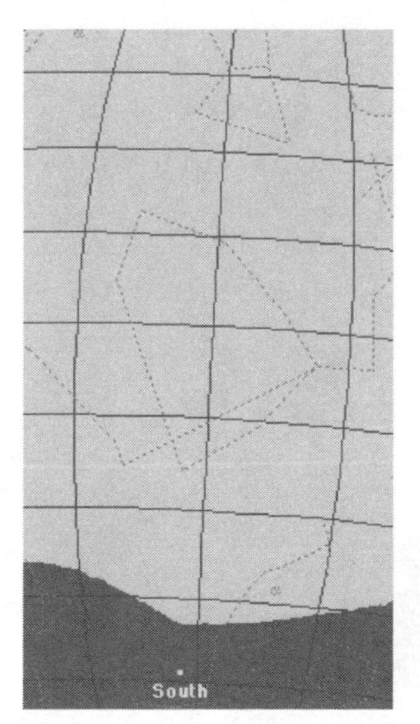

Figure 4-3 Draw each analemma that you observed.

What difference do you see between the analemmas?

Can you think of a reason for this difference?

PARTS 3 AND 4: AROUND THE WORLD (TWICE)

Record your observations of the transit altitude and hours that the Sun was in the sky at each latitude on the different dates. Use your observations to answer the questions below.

Latitude	June 21 Transit Altitude	Dec. 21 Transit Altitude	June 21 Hours in Sky	Dec. 21 Hours in Sky
Far North				
Home*				
Equatorial				
Southern Location				
Far South				

*You can transfer your results from Part 1 for your home latitude.

What happens to the Sun's altitude when you travel toward the poles?

Toward the equator?

What happens to the seasonal variation in the number of hours that the Sun is in the sky as you travel toward the poles?

Toward the equator?

Based on your data table, which is more important for warm and cold weather, the amount of time the Sun is in the sky or its altitude? Justify your answer from your data. *Hint:* Concentrate on the data for the poles and the equator.

Compare data for your home latitude and the comparable southern latitude. What difference do you see in seasonal variations in the southern hemisphere compared to those in the northern hemisphere? *Hint:* This could be answered in one word.

CONCLUSION

In the space below, write a conclusion for this activity. Briefly explain what you did and what you learned from it.

CHECK YOUR UNDERSTANDING 4: MOTIONS OF THE SUN

MULTIPLE-CHOICE QUESTIONS

1. In which location would there be a day where the Sun is up for 24 hours straight?
   a. a middle-north latitude
   b. the equator
   c. the North Pole
   d. [that can never be observed]
2. In which location would the day with the least sunlight be Dec. 21?
   a. a middle-north latitude
   b. the equator
   c. the South Pole
   d. [all the locations have the least sunlight on that date]
3. In which location could the Sun be observed at the zenith?
   a. a middle-north latitude
   b. the equator
   c. the North Pole
   d. [none of the above]
4. Which location has the least variation in the daily amount of sunlight throughout the year?
   a. a middle-north latitude
   b. the equator
   c. the North Pole
   d. the South Pole
5. On which date would a location in Australia observe the midday Sun at its highest altitude for the year?
   a. March 21
   b. June 21
   c. September 21
   d. December 21
6. On which date would an observer on the North Pole not see the Sun at all?
   a. March 21
   b. June 21
   c. September 21
   d. December 21

7. What would an observer on the South Pole observe on March 21?
   a. 24 hours of sunlight
   b. 24 hours of darkness
   c. sunrise
   d. sunset
8. On which date would the Sun appear its lowest when observed from the equator?
   a. March 21
   b. June 21
   c. September 21
   d. [none of the above]
9. On which date would the Sun appear its highest when observed from the South Pole?
   a. March 21
   b. June 21
   c. September 21
   d. [none of the above]
10. Which location(s) gets the most hours of daylight over the entire year?
    a. the poles
    b. the equator
    c. midlatitudes
    d. [throughout a whole year, all locations receive the same amount of sunlight]

OPEN-ENDED ACTIVITY

Use the skills you learned in this activity to gather the same data you recorded for the table in **Parts 3** and **4** for *March 21* and *September 21*. Construct a table similar to the one from **Parts 3** and **4** for *March 21* and *September 21* and plot graphs of both *Transit Altitude* and *Hours in the Sky* as a function of all four dates: *March 21*, *June 21*, *September 21*, and *December 21*. Plot a line on each graph for all five locations shown on the tables.

5

The Ecliptic

INTRODUCTION

In this activity, you will add a path for the Sun to the celestial sphere model. The Sun's path through the stars, known as the *ecliptic*, will allow you to use the celestial sphere to predict the motion of the Sun as it will be observed at different times throughout the year from various locations on Earth. The Sun's motion along the ecliptic also will allow you to determine which stars will be visible at night from season to season and understand why these changes occur. Read about the ecliptic on pages 5–6 of *Appendix A: Basic Concepts*.

PART 1: THROUGH THE YEAR

File | Open Settings | Basics | Three Cities

1. Close all the windows except the one titled **Sydney** and the **Time Panel.** Select a time step of **1 day** and expand the **Time Panel**.
2. From under the **Control** menu, bring up the **Location Panel** and move the **location marker** ⊕ to your approximate location.
3. Set the time to March 21 of this year at noon.
4. Under the **Chart** menu select **Sky View|Star Atlas Coordinates**. Also select **Chart Labels** and deselect **Star Names**.
5. Under the **Display** menu, select **Reference Markers**, then **Celestial Equator**. Repeat this to enable **Celestial Poles**, **Ecliptic Equator**, **Equinoxes**, and **Solstices**. Deselect (remove the checkmark from) all the others.
6. Finally, click on the **planet** button, the top button of the menu bar above the up and down scroll bar, to turn on the images of the planets.
7. Bring up the **Planet Panel** under the **Control** menu. Select the Sun and deselect all the other planets. Center on and lock the Sun by double-clicking its box and highlighting the lock icon.
8. Your **Planet Panel** should look similar to Figure 5-1.

Planet Panel

Planets / Comets / Asteroids / Satellites

Planets		Name	Sym	Orbit	Path	Trail
Sun		X				
Mercury		X	X			
Venus		X	X			
Earth ⊕			X			
Mars		X	X			
Jupiter		X	X			
Saturn		X	X			
Uranus		X	X			
Neptune		X	X			
Pluto		X	X			
Moon		X	X			
Shadow						

Figure 5-1 The **Planet Panel** with the Sun selected and locked

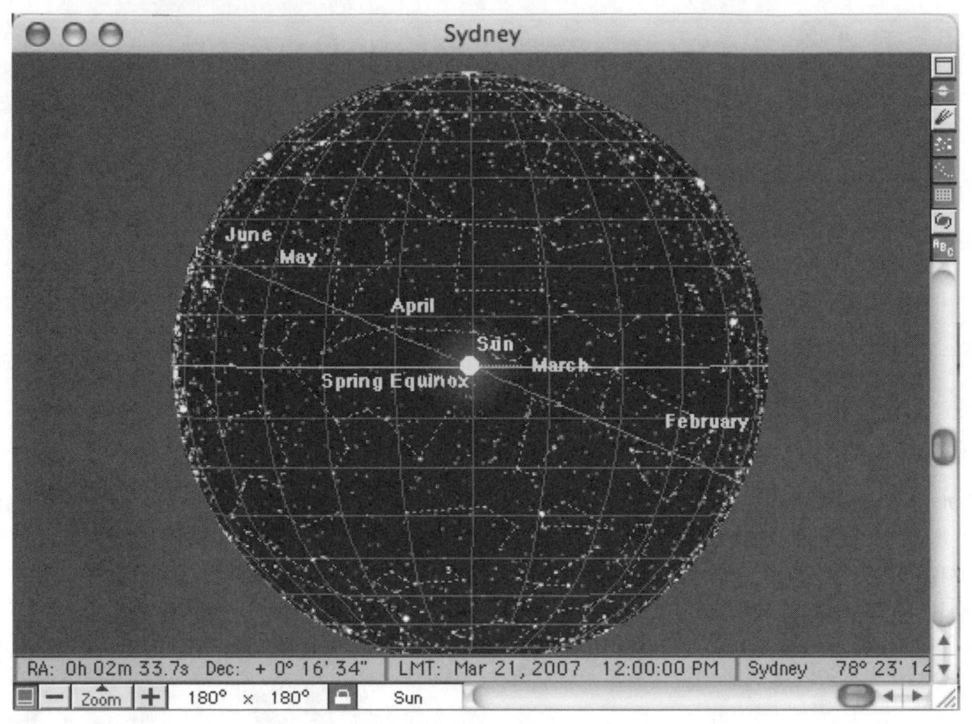

Figure 5-2 Sky Chart showing the Sun and the ecliptic

9. Use the **Zoom** buttons and the scroll bars or arrow keys to adjust your view until your Sky Chart looks similar to Figure 5-2.
10. Advance the time in 1-day increments and observe the Sun's motion on the ecliptic. Stop at the dates on the RESULTS sheet and record the information requested.
11. Use your data to answer the questions below the data table on the RESULTS sheet.
12. Use the **Location Panel** to move to a southern hemisphere location with a southern latitude comparable to your northern latitude. Set the time to June 21 at noon and answer the last question in Part 1 of the RESULTS sheet.

PART 2: THE SUN OVER THE EARTH

1. Set your location to Detroit, Michigan, using the **List Cities** button in **Set Location**.
2. Advance the time, beginning at March 21, and watch the Sun move around the ecliptic. Also, pay special attention to the **Location Panel** and observe the changing position of the Sun over the Earth. Follow the instructions and answer the questions on the RESULTS sheet.

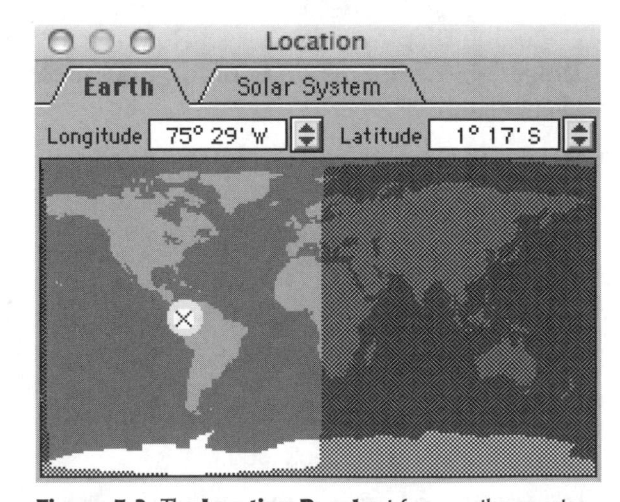

Figure 5-3 The **Location Panel** set for near the equator on March 21

PART 3: THE SUN OVERHEAD

1. Set the date to March 21.
2. This time, drag the **location marker** ⊕ on the **Location Panel** and center it on the Sun. This will tell you the Sun's latitude on that date. See Figure 5-3.
3. Also do this for June 21, September 21, and December 21. Record your data and answer the

questions on the RESULTS sheet.

PART 4: WHO ORBITS WHOM?

1. Select **Paths of the Planets** under the **Explore** menu. Then select **Sun Along the Ecliptic**. See Figure 5-4.
2. Advance the time to watch the **View from above the Sun**. Also watch the **View from Earth**. See Figure 5-4. Answer the questions on the RESULTS sheet. *Hint:* As the Earth rotates, it is noon at the point facing the Sun, and midnight at the point facing away from the Sun. Keep this in mind when you answer the questions on the RESULTS sheet.

This is a bit of a warm-up for the next activity.

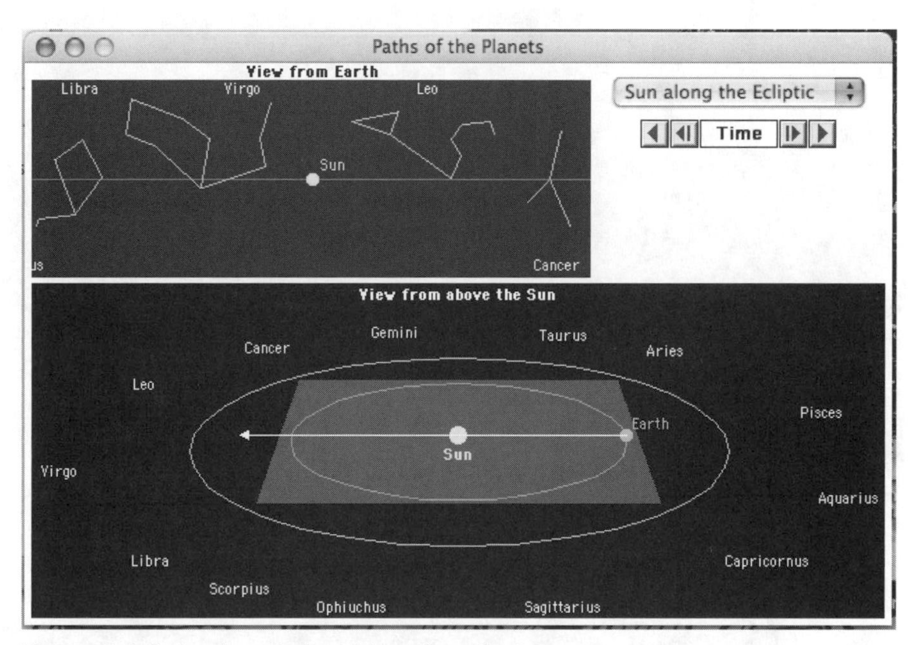

Figure 5-4 A view from Earth and from above the Sun

RESULTS SHEET 5 The Ecliptic

NAME ____________________ **DATE** ____________ **SECTION** ____________________

PART 1: THROUGH THE YEAR

Record the information from the Sun's **Data Panel** for each date on the table below. **Transit** altitude is found under the **Visibility** tab. All the other information you need is found under the **General** tab. Use the information you recorded on the table to answer the questions below.

Date	Right Ascension	Declination	Constellation	Transit Altitude
March 21				
June 21				
Sept 21				
Dec 21				

During which season(s) is the Sun's declination highest?

During which season(s) is it lowest?

How does the variation in declination compare to the variations in the Sun's altitude (viewed from your location) during those seasons?

The Sun's declination is INCREASING | DECREASING *(circle one)* from June to December.

The Sun's declination is INCREASING | DECREASING *(circle one)* from December to June.

On what date(s) does the Sun cross the celestial equator?

What is the angle between the ecliptic and the celestial equator?

When you changed to a southern hemisphere location of comparable southern latitude to your northern latitude location, the Sun's DECLINATION | TRANSIT ALTITUDE *(circle one)* changed. The DECLINATION | TRANSIT ALTITUDE *(circle one)* stayed the same. What does this tell you about the seasons in the southern hemisphere?

PART 2: THE SUN OVER THE EARTH

Figure 5-5 shows the **Location Panel** for March 21, with the **location marker** ⊕ on Detroit, Michigan. Label the Sun with the date March 21; draw the Sun at its location on June 21, September 21, and December 21; then answer the questions below.

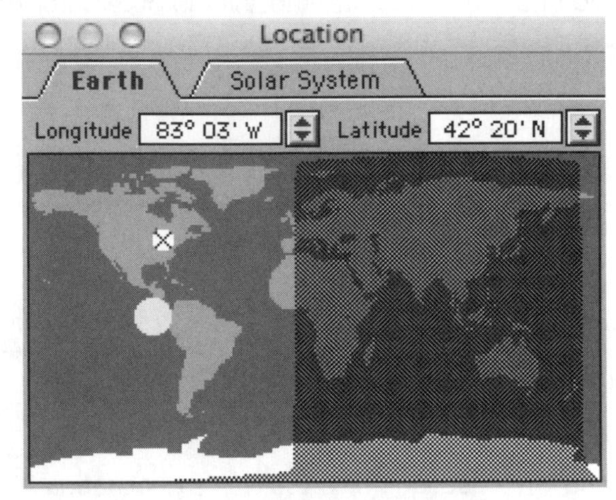

Figure 5-5 The **Location Panel** set for Detroit, Michigan on March 21

On which date is the Sun closest to being over Detroit?

On which date is the Sun farthest from being over Detroit?

The Sun is moving toward Detroit from (dates) ________________ to ________________.

The Sun is moving away from Detroit from (dates) ________________ to ________________.

How would these answers differ for Wellington, New Zealand? Try it.

PART 3: THE SUN OVERHEAD

For each date, record the latitude the Sun is directly overhead on the **Location Panel**. Use the information you recorded on the table to answer the questions below.

Date	Latitude the Sun Is Directly Over
March 21	
June 21	
September 21	
December 21	

How do the latitudes in the above table compare with the declinations in the table of Part 1?

If the Sun is directly over a latitude, what would the observed altitude of the Sun be at that location?

Would the Sun ever appear overhead in Detroit? Explain why or why not.

At your location (if not Detroit)? Explain why or why not.

PART 4: WHO ORBITS WHOM?

As ____________ orbits the ____________, the ____________ appears to move through the stars on the ecliptic.

How long will the apparent motion around this path take?

When the Sun is in Capricorn, what constellation would appear high overhead at midnight to an observer on Earth? Explain why this would be the case.

When Taurus appears high overhead at midnight to an observer on Earth, what constellation is the Sun in?

These constellations on and around the ecliptic, those that the Sun appears to move through, are collectively known as the ___________________.

CONCLUSION

In the space below, write a conclusion for this activity. Briefly explain what you did and what you learned from it.

CHECK YOUR UNDERSTANDING 5: THE ECLIPTIC

MULTIPLE-CHOICE QUESTIONS

1. On which date does the Sun reach the northernmost point in declination on the ecliptic?

    a. March 21

    b. June 21

    c. September 21

    d. December 21

2. The points where the ecliptic and the celestial equator intersect are called

    a. cardinal points.

    b. celestial poles.

    c. equinoxes.

    d. solstices.

3. On which date is the Sun directly over the equator?

    a. March 21

    b. June 21

    c. September 21

    d. [two of the above]

4. December 21 is the date that the Sun is over

    a. its farthest north latitude.

    b. its farthest south latitude.

    c. the equator.

    d. the zenith.

5. June 21 is the date that the Sun is

    a. at its northern-most declination.

    b. at its southern-most declination.

    c. at its highest altitude observed in the southern hemisphere.

    d. [more than one of the above]

6. On which date will the Sun be observed at it highest midday altitude in the southern hemisphere?

    a. December 21

    b. June 21

    c. September 21

    d. [two of the above]

7. Observed from the northern hemisphere between March 21 and June 21, the Sun's __________ will be increasing.

   a. declination

   b. midday altitude

   c. [both will be increasing]

   d. [neither will be increasing]

8. Observed from the southern hemisphere between March 21 and June 21, the Sun's declination and midday altitude will be

   a. both increasing.

   b. both decreasing.

   c. increasing and decreasing, respectively.

   d. decreasing and increasing, respectively.

9. The Sun appears to move all the way around the ecliptic once per

   a. day.

   b. week.

   c. month.

   d. year.

10. About how many hours of a right ascension will the Sun move in a month?

    a. 1

    b. 2

    c. 12

    d. 24

OPEN-ENDED ACTIVITY

Use the skills you learned in this activity to determine approximately the month the Sun is within the borders of each zodiac constellation.

VOYAGER: SKYGAZER CD-ROM ACTIVITIES

6

Seasonal Constellations

INTRODUCTION

As Earth orbits the Sun, the Sun appears to move through the sky along a path we call the *ecliptic*. This causes the stars that are visible at night to change from season to season throughout the year. In this activity you will learn which stars and constellations to look for at different times of the year.

PART 1: VIEWING ZODIACAL CONSTELLATIONS

1. Select **Paths of the Planets** under the **Explore** menu. Then select **Sun Along the Ecliptic**.
2. Advance the time to watch the **View from above the Sun**. Also watch the **View from Earth**. Keep in mind that as Earth rotates, it is noon at the point facing the Sun and midnight at the point facing away from the Sun. The points halfway between are sunrise and sunset. Also remember that Earth rotates from west to east, which is counterclockwise when viewed from above. Fill in the table on the RESULTS sheet.

PART 2: SEASONAL STARS AND CONSTELLATIONS

1. Select **Season Sky Calendar** under the **Explore** menu.
2. Click on the "ABC" button to enable the names of the stars and constellations.
3. Select the month **May**. Your **Star Calendar** should look similar to Figure 6-1.
4. The **Star Calendar** shows you which stars and constellations are visible and where they are in the sky at different times of year. This particular **Star Calendar** is valid for observers from middle-north latitudes, where most people, probably including you, live. When using a **Star Calendar** the direction that the observer is facing must be placed

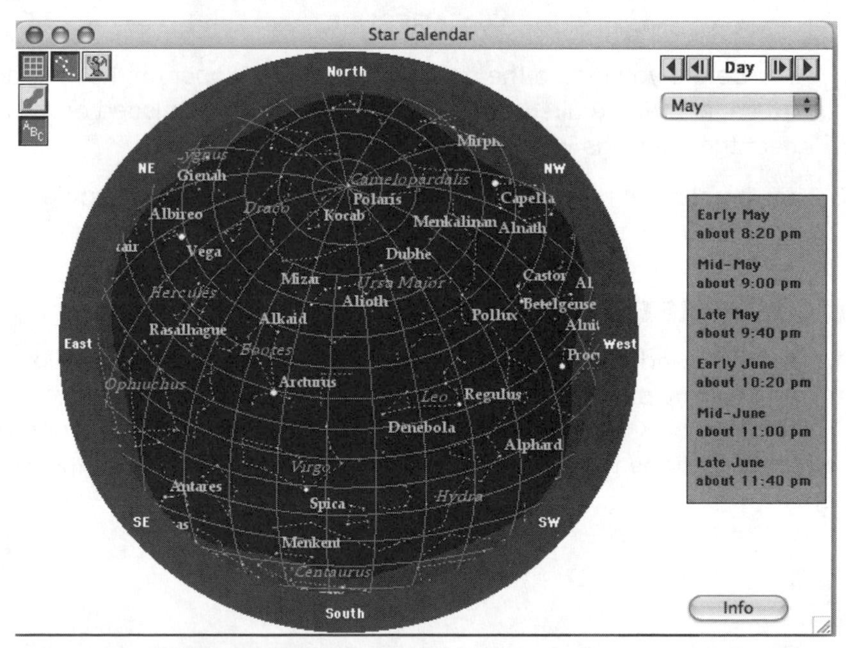

Figure 6-1 The **Star Calendar** can be used to determine the times and locations that stars and constellations will be visible

downward, so your **Star Calendar** is oriented for an observer facing south. Record the names of some of the constellations you see high in the southern sky and their brightest stars on the RESULTS sheet. May is part of the season spring, so these are spring constellations.

5. Record the names of some of the constellations you see high in the southern sky and their brightest stars on the RESULTS sheet for August (summer), November (fall), and February (winter).
6. You can make your **Star Calendar** much larger and easier to see by clicking on the green button, the rightmost of the three buttons in the upper-left corner. You can adjust the **Star Calendar** to any size by clicking and dragging the bottom right-corner.

PART 3: WHAT'S UP AND WHEN?

1. Return to the month **May** on your **Star Calendar** and this time look at the stars and constellations that are rising in the east and those that are setting in the west.
2. Record on the RESULTS sheet which of the four *seasonal groups* (spring, summer, fall or winter, identified in PART 2) of constellations you see rising in the east. Also record which group is seen high in the south, which group is seen in the west, and finally which group you do not see at all.
3. Repeat this procedure for August (summer), November (fall), and February (winter). For each time, record on the RESULTS sheet which *seasonal groups* of constellations you see where at which times. If you think you are seeing a pattern, try to predict where each *seasonal group* will be visible and at what times before you check your answers.

PART 4: CIRCUMPOLAR CONSTELLATIONS

1. Observe your **Star Calendar** during each month of year and record on the RESULTS sheet the names of several stars and constellations that can be seen all year around. Note in which part of the sky you see them.
2. Note the position of the Big Dipper, the outlined part of *Ursa Major*. Observe the position of the Big Dipper in May, June and July. Record the position of the Big Dipper in May on your RESULTS sheet and then, using the motion you have observed, predict the positions of the Big Dipper in August. Record your prediction on your RESULTS sheet and check to see if it is correct.

   Hint: Before you record the position of the Big Dipper on the diagrams on your RESULTS sheet, remember that the direction the observer is facing is placed *downward* on a **Star Calendar**. This will affect the drawings you make on your RESULTS sheet.
3. Now predict the positions of the Big Dipper in November and February. Record your predictions on your RESULTS sheet and check to see if they are correct.

PART 5: MORE CONSTELLATIONS

1. The command **The Constellations** under the **View** menu will allow you to select and center your chart on any of the 88 official constellations. The **Data Panels** that come up will include information on the visibility of the constellation. Select **Star Atlas View** under the **Chart** menu and use **The Constellations** to fill out the requested data on the table on the RESULTS sheet.

RESULTS SHEET 6 Seasonal Constellations

NAME ______________________ **DATE** ______________ **SECTION** ______________________

PART 1: VIEWING ZODIACAL CONSTELLATIONS

In the table below, name the zodiacal constellations that can be seen at each of the times when the Sun is in each of the constellations in the table below. Remember that it is noon at the point on Earth facing the Sun, midnight at the point on Earth facing away from the Sun and sunrise and sunset at the points in between. Also, Earth rotates from west to east, which is counterclockwise when viewed from above the north-pole.

If you think you are seeing a pattern, try to predict, before you check, which constellations will be visible.

Sun Is in	Setting in the West at Midnight	High in the Southern Sky at Midnight	Rising in the East at Midnight
Pisces			
Gemini			
Virgo			
Sagittarius			

Were your predictions of the pattern accurate?

PART 2: SEASONAL STARS AND CONSTELLATIONS

For each season in the table below, record the names of three or four of the constellations you can see high in the south and some of their bright stars. The constellations seen high in the south during a given season are part of that season's *seasonal group*.

If you want star charts to use for star gazing, print out your Star Calendar for any month you choose.

File | Print Star Calendar

Season (Month)	Major Constellations	Bright Stars
Spring (May)		
Summer (August)		
Fall (February)		
Winter (November)		

PART 3: WHAT'S UP AND WHEN?

Record which of the four *seasonal groups* (the constellations of spring, summer, fall, or winter) you saw (or did not see) in each position on each date.

Month	Seasonal Group Seen Rising in the East	Seasonal Group Seen High in the South	Seasonal Group Seen Setting in the West	Seasonal Group Not Seen
May				
August				
November				
February				

During each season, which seasonal group of constellations do you not see and why?

PART 4: CIRCUMPOLAR CONSTELLATIONS

Constellations seen all-year-around:

Stars seen all-year-around:

Direction you look to see these stars and constellations:

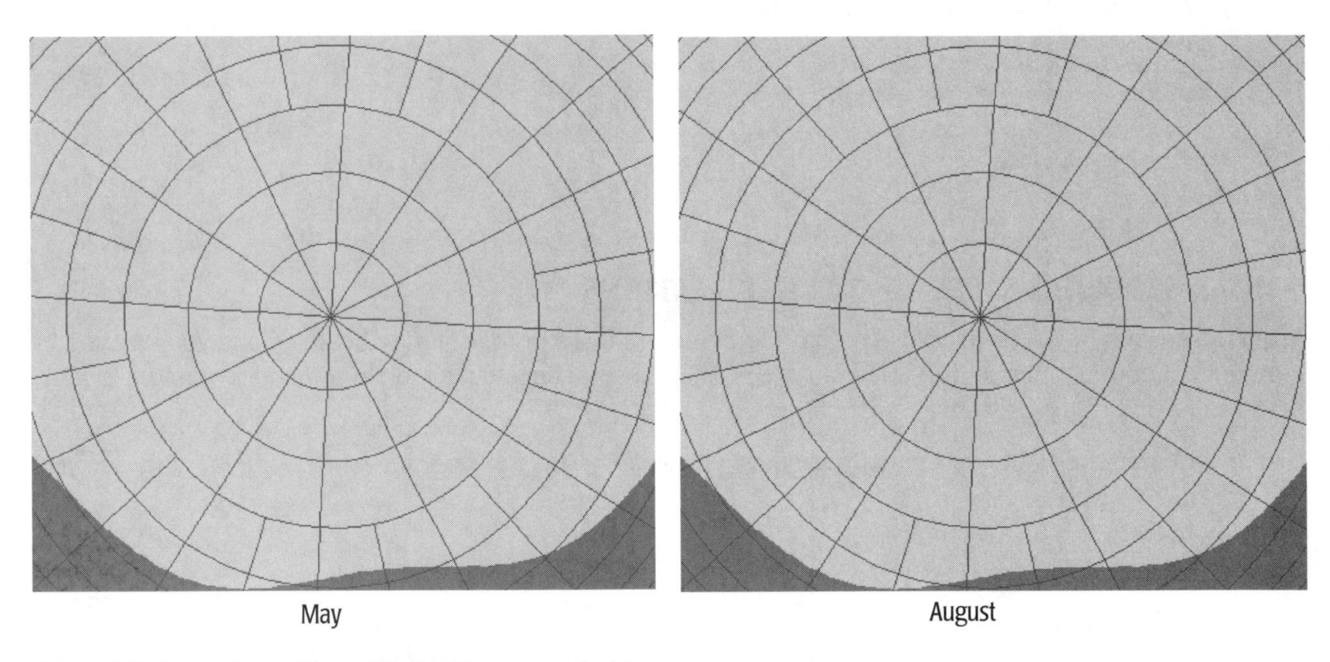

Figure 6-2 Record the positions of the Big Dipper on each date

On the left panel of Figure 6-2, record the approximate position of the Big Dipper in the Northern sky in May.

On the right panel, record your prediction for the position of the Big Dipper in August.

Was your prediction correct?

If not, record the correct position as well.

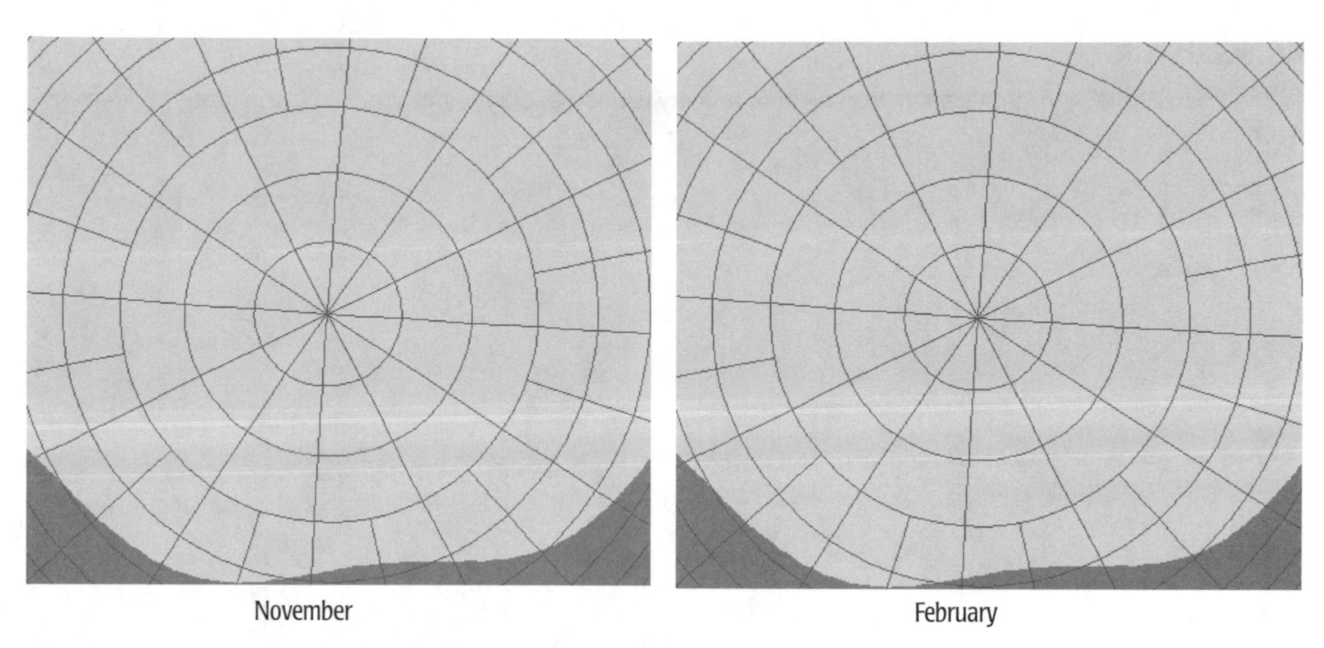

Figure 6-3 Record the positions of the Big Dipper on each date.

On Figure 6-3, record your predictions of the approximate positions of the Big Dipper in November (left panel) and February (right panel).

Were your predictions correct?

If not, record the correct positions as well.

PART 5: MORE CONSTELLATIONS

Fill out the information about each constellation in the table below.

Constellation	Brightest (α) Star (if labeled)	Midnight Transit	Seasonal group*	Visible from Your Location?
Bootes				
Lyra				
Pegasus				
Orion				
Cassiopeia				
Centaurus				

*Does the midnight transit of each constellation place it in the same seasonal group as in Part 2?

CONCLUSION

In the space below, write a conclusion for this activity. Briefly explain what you did and what you learned from it.

CHECK YOUR UNDERSTANDING 6: SEASONAL CONSTELLATIONS

MULTIPLE-CHOICE QUESTIONS

1. Which constellation would be visible any night of the year from the northern hemisphere?
   a. Gemini
   b. Virgo
   c. Sagittarius
   d. Ursa Major
2. Which seasonal group of constellations will be rising in the east March 21 at midnight?
   a. spring
   b. summer
   c. winter
   d. fall
3. What season is it if the summer seasonal group of constellations is setting in the west at sunrise?
   a. spring
   b. summer
   c. winter
   d. fall
4. What time is it if the fall seasonal group of constellations is high overhead in October?
   a. noon
   b. sunset
   c. midnight
   d. sunrise
5. What season is it if you stay up all night and do not see the winter seasonal group of constellations at all?
   a. spring
   b. summer
   c. winter
   d. fall
6. Which constellation is high overhead at midnight in the winter?
   a. Pisces
   b. Gemini
   c. Virgo
   d. Sagittarius

7. Which constellation is rising in the east at midnight in the spring?
   a. Pisces
   b. Gemini
   c. Virgo
   d. Sagittarius
8. Which constellation is setting in the west at midnight in the summer?
   a. Pisces
   b. Gemini
   c. Virgo
   d. Sagittarius
9. If Virgo is rising in the east at midnight, it is
   a. fall.
   b. winter.
   c. spring.
   d. summer.
10. If Sagittarius is high in the southern sky at midnight, it is
    a. fall.
    b. winter.
    c. spring.
    d. summer.

OPEN-ENDED ACTIVITY

Use the skills you learned in this activity to find and name one constellation that is currently each of the following: rising in east, high in the south, setting in the west, and not visible at all from your location at midnight. Determine the seasonal group of which each constellation you name is a part. Which constellation is part of the group seen during the current season?

VOYAGER: SKYGAZER CD-ROM ACTIVITIES

7

The Seasons

INTRODUCTION

In Activity 5, you observed the variations in the Sun's midday altitude and the hours that it is up in the sky. Now you will investigate the reasons for these observed changes. Read about the seasons on pages 6–7 of *Appendix A: Basic Concepts*; also see Astronomy Place Tutorial 2 for more about the seasons.

PART 1: THE VIEW FROM SPACE

1. Select **Seasons of the Earth** under the **Explore** menu. Select **Summer** and observe Earth's tilt.
2. Advance time to rotate Earth and observe that different latitudes spend different amounts of time in daylight and darkness. Answer the questions on the RESULTS sheet.
3. Now select **Winter**.
4. Advance time and make the same observations as you did for summer. Answer the questions on the RESULTS sheet.
5. Advance time and allow Earth to revolve around the Sun. Observe the gradual changes between the seasons. Can you see why the Sun is both up and down for 12 hours on the first days of spring and fall? *Equinox* is Latin for equal night.
6. Click on the "info" button in the bottom-right corner and read about the seasons.

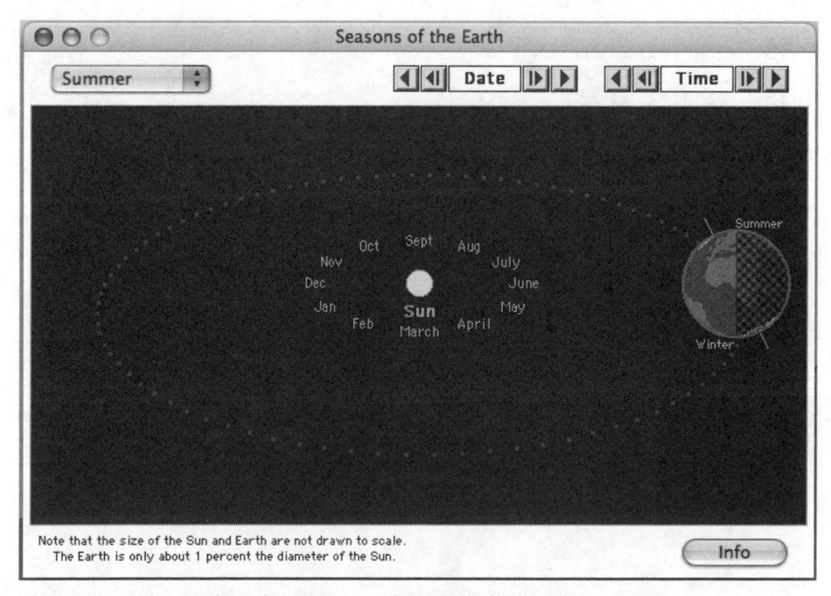

Figure 7-1 The Earth with its rotational axis tilted toward the Sun

PART 2: THE VIEW FROM THE SUN

File | Open Settings | Basics | Earth's Seasons

In this setting you are observing Earth from the position of the Sun.

1. Advance the time and view the changes that occur throughout a year.
2. Answer the questions on the RESULTS sheet.

PART 3: SEASONS ON OTHER PLANETS

1. Using the same setting as in Part 2, select **Center Planet** under the **View** menu and choose **Mars**.
2. Zoom in until Mars appears as large as Earth did.
3. Now advance the time in 2-day increments. Pay special attention to Mars's polar ice caps. Does Mars appear to have seasonal changes similar to those on the Earth?
4. Select **Venus** from **Center Planet** under the **View** menu. You may have to zoom out, since Venus is closer to the Sun than Mars. Does Venus appear to have seasonal changes? What do you notice about Venus's rotation that is different than Earth's and Mars's?
5. Now select **Uranus**.
6. Zoom in and compare and contrast the rotation with that of the Earth and the other planets you have observed. Answer the questions on the RESULTS sheet.

PART 4: LOOKING AT THE REST OF THEM

Using the same commands as in Part 3 (but with smaller time steps), observe the rotations of Jupiter, Saturn, and Neptune, then answer the questions on the RESULTS sheet.

PART 5: MORE VIEWS FROM SPACE

File | Open Settings | Earth's Seasons 2
File | Open Settings | Earth's Seasons 3

Use these settings to observe Earth's seasons again, but from different perspectives.

RESULTS SHEET 7 The Seasons

NAME ______________________ **DATE** ______________ **SECTION** ______________________

PART 1: THE VIEW FROM SPACE

Summer

Is your location on Earth spending more time in daylight or darkness?

How does this differ for latitudes to the north and to the south of you?

Where on Earth is it dark all day?

Light all day?

Estimate how much time is spent in daylight and how much in darkness on the equator.

Why does it say "winter" next to the South Pole?

Winter

Is your location on Earth spending more time in daylight or darkness?

How does this differ for locations to the north and to the south of you?

Where on Earth is it dark all day?

Light all day?

Estimate how much time is spent in daylight and darkness on the equator.

Why does it say "summer" next to the South Pole?

Spring and Fall

What is the reason for our use of the term *equinox*?

PART 2: THE VIEW FROM THE SUN

Match the correct date with each occurrence. (Some dates may be used more than once and an event can occur on more than one day.)

A = March 21 B = June 21 C = September 21 D = December 21

_____________ Northern hemisphere pointed toward the Sun

_____________ Southern hemisphere pointed toward the Sun

_____________ North and South Poles equidistant from the Sun

_____________ South Pole not visible from the Sun

_____________ North Pole not visible from the Sun

_____________ Solstice

_____________ Equinox

_____________ Equator is directly under the Sun.

PART 3: SEASONS ON OTHER PLANETS

For each planet you observed, draw its axis of rotation in Figure 7-2, showing its tilt relative to Earth's tilt.

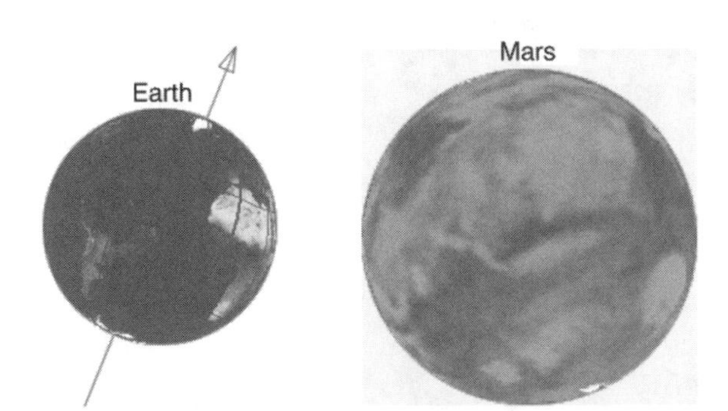

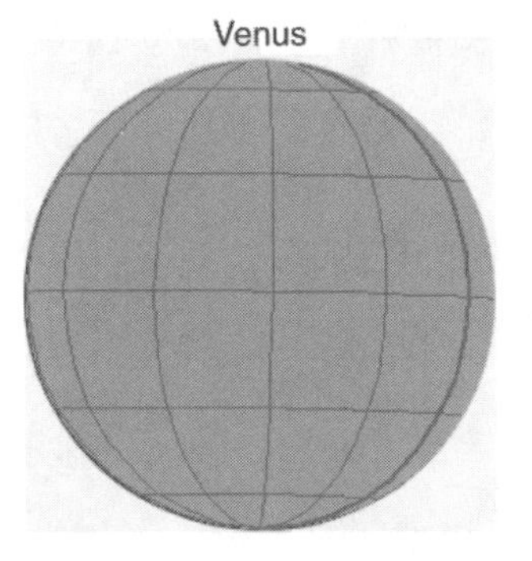

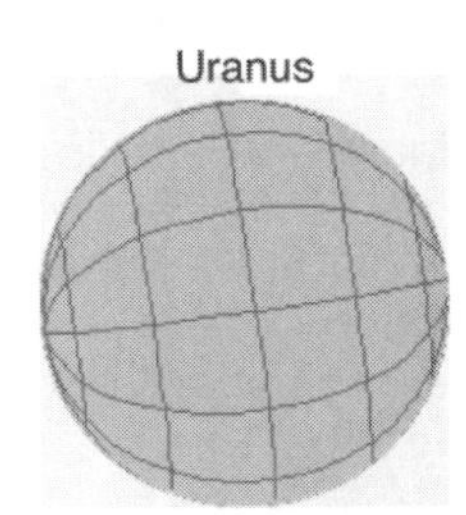

Figure 7-2 Draw the axis of rotation on each planet

Based on your drawings, answer the following questions:

Which planet has seasons most similar to Earth's? Explain your answer.

Which planet has no seasons at all? Explain your answer.

Which planet experiences seasons similar to those on Earth's poles for its entire year? Explain your answer.

What is different about Venus's rotation compared to the other planets you observed?

PART 4: LOOKING AT THE REST OF THEM

Was Jupiter's tilt closest to that of Earth, Venus, or Uranus?

Was Saturn's tilt most similar to that of Earth, Venus, or Uranus?

If Neptune experienced seasons in the same way that Earth does, which season would it be on Neptune? Explain your answer. Remember, the illuminated side of Neptune is facing the Sun.

CONCLUSION

In the space below, write a conclusion for this activity. Briefly explain what you did and what you learned from it.

CHECK YOUR UNDERSTANDING 7: THE SEASONS

MULTIPLE-CHOICE QUESTIONS

1. Which planet has the most axial tilt?
   a. Earth
   b. Mars
   c. Jupiter
   d. Uranus
2. Which planet should have seasons most similar to Earth's?
   a. Venus
   b. Mars
   c. Jupiter
   d. Uranus
3. Which planet could be thought of as "upside down" compared to the others?
   a. Venus.
   b. Earth.
   c. Mars.
   d. Jupiter.
4. Earth's northern hemisphere is tilted toward the Sun in
   a. March.
   b. June.
   c. September.
   d. December.
5. When Earth's northern hemisphere is tilted toward the Sun, it is
   a. summer.
   b. winter.
   c. summer in the northern hemisphere and winter in the southern hemisphere.
   d. winter in the northern hemisphere and summer in the southern hemisphere.
6. Neither of Earth's hemispheres is tilted closer to the Sun in
   a. March.
   b. June.
   c. September.
   d. [two of the above].

7. An equinox occurs
   a. halfway between solstices.
   b. when the Earth's North and South Poles are equally distant from the Sun.
   c. all over the Earth at the same time.
   d. [all of the above].
8. A solstice occurs
   a. in June and December.
   b. when one of the poles is tilted toward the Sun.
   c. in between the equinoxes.
   d. [all of the above].
9. Extremes in the amounts of daylight and darkness occur at the
   a. solstices.
   b. equinoxes.
   c. [both of the above depending on which hemisphere you are in].
   d. [none of the above].
10. Which planet would experience the least seasonal changes throughout its orbit around the Sun?
    a. Earth
    b. Mars
    c. Jupiter
    d. Uranus

OPEN-ENDED ACTIVITY

Use the skills you learned in this activity and the previous one to devise an experiment that will prove that the altitude of the Sun is more important than the number of hours that it is up in determining whether the weather and climate at a given location will be warmer or colder.

8

Precession

INTRODUCTION

Earth rotates on its axis once per day and revolves around the Sun once per year. *Precession* is the conical-shaped wobble of Earth's axis over time. Precession is the change in the direction of the tilt, *not* the amount of tilt. Precession takes 26,000 years, but apparent changes in the sky that are a result of precession are noticeable in considerably less time.

Throughout this activity you can see an example of the real power of *Voyager: SkyGazer*. You will be able to simulate changes in the sky due to precession that you would never be able to observe in a lifetime.

PART 1: CHANGING THE POLE STAR

1. Select **Wobble of the Earth** under the **Explore** menu.
2. Click on the **Time** button to observe Earth rotating about its axis.
3. Click on the **Year** button to observe Earth's precession.
4. You can run the year backward and forward and observe precession's effect on the polar region of the sky. Use these buttons to answer the questions on the RESULTS sheet.
5. Click on the "**Info**" button in the bottom right corner and read about precession.

PART 2: POLARIS'S DRIFT

File | Open Settings | Basics | North Pole 3

1. Advance the time in 2-year steps to observe the apparent motion of Polaris due to precession over the next 500 years.
2. Record Polaris's path and answer the questions on the RESULTS sheet.

File | Open Settings | Settings | Polaris Nears the Pole

(Up to this point, most settings you have run have been from the **Basics** folder. In this activity, you will start using settings from the **Settings** and **Demos** folders. See Part 3 of *Activity 1 Introducing SkyGazer* on page 104 for details.)

3. Use this setting to determine when Polaris will be closest to the north celestial pole.

PART 3: POINTER STARS

File | Open Settings | Basics | Precession of the Equinoxes

1. Click on the two stars at the end of the bowl of the Big Dipper and record their names on the RESULTS sheet. These stars are known as the *pointer stars* because if you draw a line out of the Big Dipper's bowl through these two stars, it will always point to Polaris.
2. Read the **Settings File Banner** (the blue box) and note the star identified as the pole star when the Great Pyramids of Egypt were built.
3. Which two stars in the Big Dipper would have been better pointer stars in 2800 B.C.? Record your answer on the RESULTS sheet.

PART 4: PRECESSION THROUGH THE AGES

1. Use the same settings as in Part 3.
2. Click the **stars** button on the **Display Panel** to turn off the stars. This can also be done with the **stars** button on the panel above the up and down scroll bar (see Figure 8-1).

Figure 8-1 The Display Panel and the stars button

3. Now use the up and down scroll bar or arrow keys to change the declination of the Sky Chart to 0°. Use the left and right scroll bar or arrow keys to change the right ascension of the Sky Chart to 0 hrs. Your Sky Chart should look similar to the one in Figure 8-2.

The intersection point between the ecliptic and the celestial equator is the *spring* (or *vernal*) *equinox*. The Sun is found at this point in the sky 0 hrs R.A., 0° declination around March 21. We refer to the constellation in which the spring equinox appears as the name of the current *age*.

4. Click anywhere within the boundaries of the constellation that the spring equinox is in to see what age we are in.
5. Step the time both forward and backward to see the most recent and the next ages.

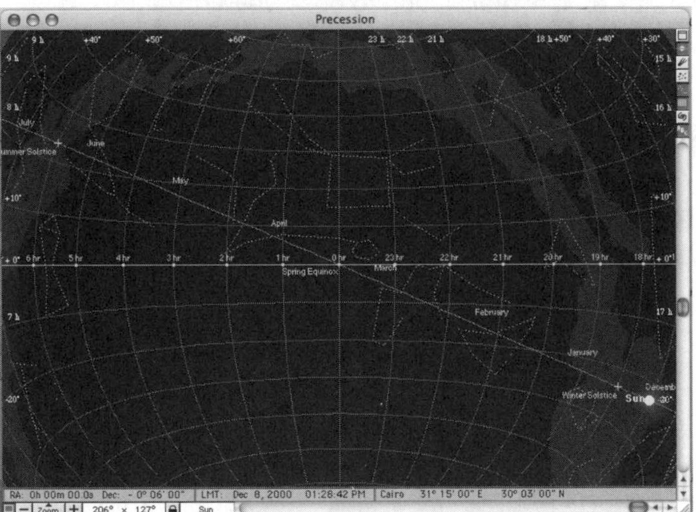

Figure 8-2 Sky Chart centered on the Spring Equinox

File | Open Settings | Settings | Age of Aquarius

Use the above setting to check your answers.

PART 5: CHANGES

File | Open Settings | Basics | Chicago 10,000 AD

1. Answer the following questions on your RESULTS sheet: Which star will be the pole star in 10,000 A.D.? Which bright winter star will no longer be seen from Chicago's latitude?

RESULTS SHEET 8 Precession

NAME ______________________ **DATE** ______________ **SECTION** ______________

PART 1: CHANGING THE POLE STAR

Which star is our current pole star?

Since when has this star been the pole star? (For this activity, the pole star is defined as the star closest to the north celestial pole of the ones shown.)

Which star was the pole star in 3000 B.C.?

What star will be the pole star 1000 years from now?

In the table below, beginning with 3000 B.C., enter the name of the then-current pole star and the next four pole stars in order. Also enter the approximate years when each one first became the pole star and stopped being the pole star. Finally, state whether you think the star was a good pole star or not, based on how close was it to the north celestial pole. Again, consider only the stars that are labeled.

Star	Year Became Pole Star	Year Stopped Being Pole Star	Was It a Good Pole Star?

Which stars do you think are/were the best and worst pole stars?

Since precession changes the direction that the Earth's rotational axis is pointing in space, what other observable effect besides the changing of the North Star would you expect precession might have?

Hint: What have you already studied that is caused by the tilt of the Earth's rotational axis?

PART 2: POLARIS'S DRIFT

Starting at the present, record the motion of Polaris due to precession over the next few hundred years. In Figure 8-3, label its position every 50 years, starting in 2000 A.D.

Is Polaris ever exactly on the north celestial pole?

If not, about when will it be closest?

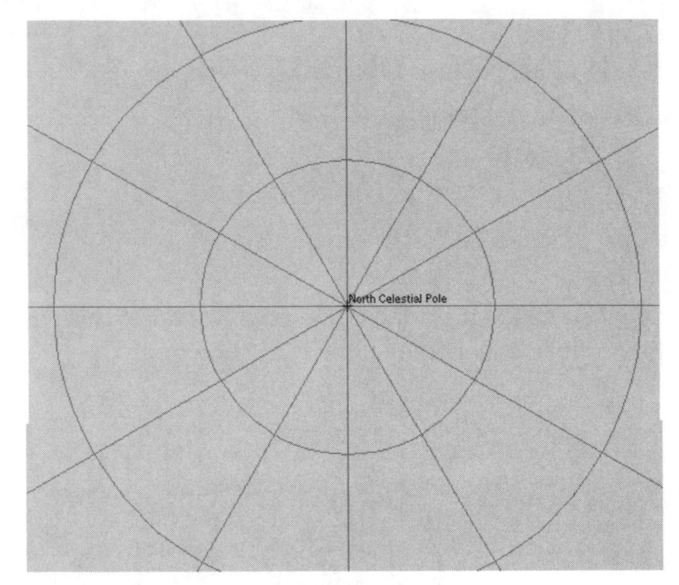

Figure 8-3 Record the future positions of Polaris

PART 3: POINTER STARS

Current pole star:

Current pointer stars:

Pole star in 2800 B.C.:

Pointer stars in 2800 B.C.:

In Figure 8-4, mark and label the position of each of these stars.

Figure 8-4 Record the positions of past and future pole stars

PART 4: PRECESSION THROUGH THE AGES

Record the current, most recent, and next ages in the table below.

Age	Constellation
Most recent age	
Current age	
Next age	

What noticeable change would precession of the equinoxes (the change of the spring equinox's position in the sky or the change in the direction of the tilt of the rotational axis) have on human society that would have to be corrected?

PART 5: CHANGES

Pole star in 10,000 A.D.:

Bright star that will no longer be visible:

CONCLUSION

In the space below, write a conclusion for this activity. Briefly explain what you did and what you learned from it.

CHECK YOUR UNDERSTANDING 8: PRECESSION

MULTIPLE-CHOICE QUESTIONS

1. The effects of precession are noticed over a time scale of

   a. years.

   b. decades.

   c. hundreds of years.

   d. thousands of years.

2. Which is not an effect of precession?

   a. changing the pole star

   b. a change in the positions of the equinoxes

   c. changes in the relative positions of stars in constellations

   d. [all of the above ARE caused by precession]

3. Which star will not be the pole star some day?

   a. Vega

   b. Deneb

   c. Altair

   d. [Polaris will always be the pole star]

4. Which star has already been the pole star?

   a. Vega

   b. Deneb

   c. Thuban

   d. [Polaris has always been the pole star]

5. Which will be the next pole star?

   a. Alderamin

   b. Vega

   c. Thuban

   d. [Polaris will always be the pole star]

6. Which will be the "best" pole star?

   a. Alderamin

   b. Deneb

   c. Vega

   d. [Polaris will always be the pole star]

7. What is meant by the "age" we are in?
   a. Which star is the pole star
   b. Which constellation the pole star is in
   c. Which constellation the spring equinox appears in
   d. [Both b and c—They change together]
8. Which age are we currently in?
   a. Aries
   b. Pisces
   c. Aquarius
   d. [none of the above]
9. How would a star chart drawn 2000 years ago differ from one drawn today?
   a. There would be a different pole star.
   b. The constellations would look different.
   c. [both of the above are true].
   d. [none of the above are true].
10. Precession is caused by
    a. the actual motion of the stars themselves.
    b. the entire sky "wobbling" around a stationary Earth.
    c. a "wobble" of the Earth's rotational axis.
    d. [none of the above].

OPEN-ENDED ACTIVITY

Use the skills you learned in this activity to determine which star was the last pole star prior to Polaris and which star will be the next pole star.

9
Proper Motion

INTRODUCTION

Proper motion is the change in a star's position relative to the background stars that is a result of the star's actual motion around the center of the galaxy. Over long periods of time, proper motion can have profound effects on our view of the sky.

Throughout this activity you can see an example of the real power of *Voyager: SkyGazer*. You will be able to simulate changes in the sky due to proper motion that you would never be able to observe in a lifetime.

PART 1: NEARBY STARS

File| Open Settings| Demos| Barnard's Star

1. Start looking at proper motion by observing Barnard's Star. First, click on the star to bring up its **Data Panel** and see how far away it is.
2. Click on some of the other stars to see how far away they are. Record your information on the RESULTS sheet. Can you figure out why the motion of Barnard's Star is noticeable in front of a fixed background?
3. Advance the time to observe the proper motion of Barnard's Star.

File| Open Settings| Basics| Tracking Proxima Centauri

4. Use the **Zoom** button to select a field of view of 8° and follow the same procedure as for Barnard's Star above. Record your observations as instructed on the RESULTS sheet.
5. Observe the proper motion of these three nearby stars, and answer the questions on the RESULTS sheet.

File| Open Settings| Settings| Motion of Wolf 359
File| Open Settings| Settings| Proper Motion
File| Open Settings| Settings| Red Dwarf Ross 154

PART 2: THE SOLAR NEIGHBORHOOD

1. Select the **Solar Neighborhood** under the **Explore** menu.
2. Select a **Distance from Sun** of 25 light-years.
3. Under **Show,** select **Bright Stars Only** and **Star Names**. Answer the questions on the RESULTS sheet.

PART 3: ONE FRAME IN A COSMIC MOVIE

File | Open Settings | Demos | Dipper + 80,000 years

1. Step the time as far back and as far forward as *Voyager: SkyGazer* will allow. Observe the effect of proper motion on the Big Dipper. Record your observations as instructed on the RESULTS sheet.

File | Open Settings | Basics | Tracking Altair

2. Follow the same procedure that you did with the Big Dipper with the constellation Aquila. Record your observations as instructed on the RESULTS sheet.

File | Open| Settings | Prehistoric Constellations

3. Advance time to the present and make note of which star exhibits the most proper motion and which constellation changes the most. Answer the questions on the RESULTS sheet.

RESULTS SHEET 9 Proper Motion

NAME ______________________ **DATE** ______________ **SECTION** ______________________

PART 1: NEARBY STARS

Draw a line on Figure 9-1 that represents the proper motion of Barnard's Star that you observed.

Distance to Barnard's Star:

Distance to some of the other stars in the field:

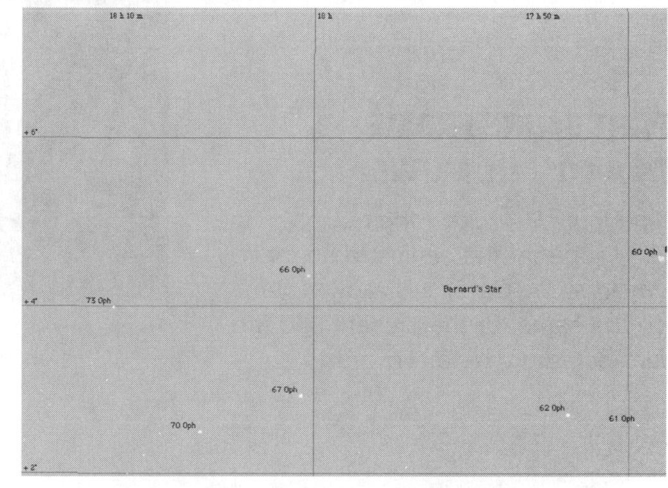

Figure 9-1 Record the proper motion of Barnard's Star

Why is the proper motion of Barnard's Star so easily detectable?

Draw a line on Figure 9-2 that represents the proper motion of Proxima Centauri that you observed.

Distance to Proxima Centauri:

Distance to some of the other stars in the field:

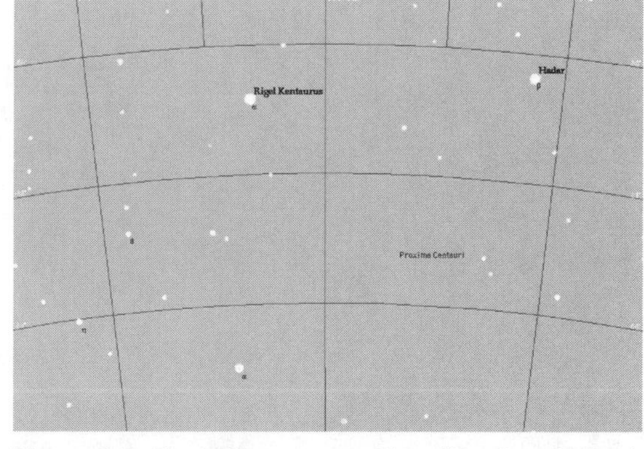

Figure 9-2 Record the proper motion of Proxima Centauri

After about 500–1000 years, what do you notice that is interesting about the nearby star Rigel Kentaurus?

Name the constellation that each star is moving through.

Wolf 359:

61 Cygni:

Ross 154:

PART 2: THE SOLAR NEIGHBORHOOD

Can you find either of the stars you tracked in Part 1? If you can, circle them on Figure 9-3.

Circle the brightest star on the figure. Is the brightest star the closest?

Circle the closest star on the figure. Is the closest star the brightest?

Explain the reason that the star that appears brightest may *not* necessarily be the closest and vice versa.

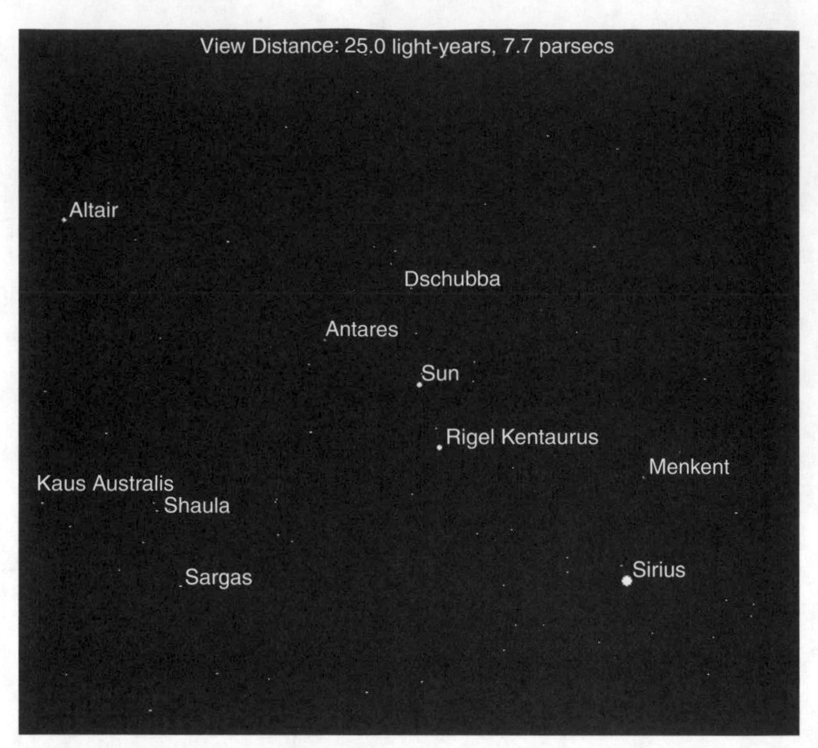

Figure 9-3 Stars in the Solar Neighborhood

PART 3: ONE FRAME IN A COSMIC MOVIE

On Figure 9-4, draw the outline of the Big Dipper at the earliest date that *Voyager: SkyGazer* will allow. Then do the same for the present and the furthest future date allowed.

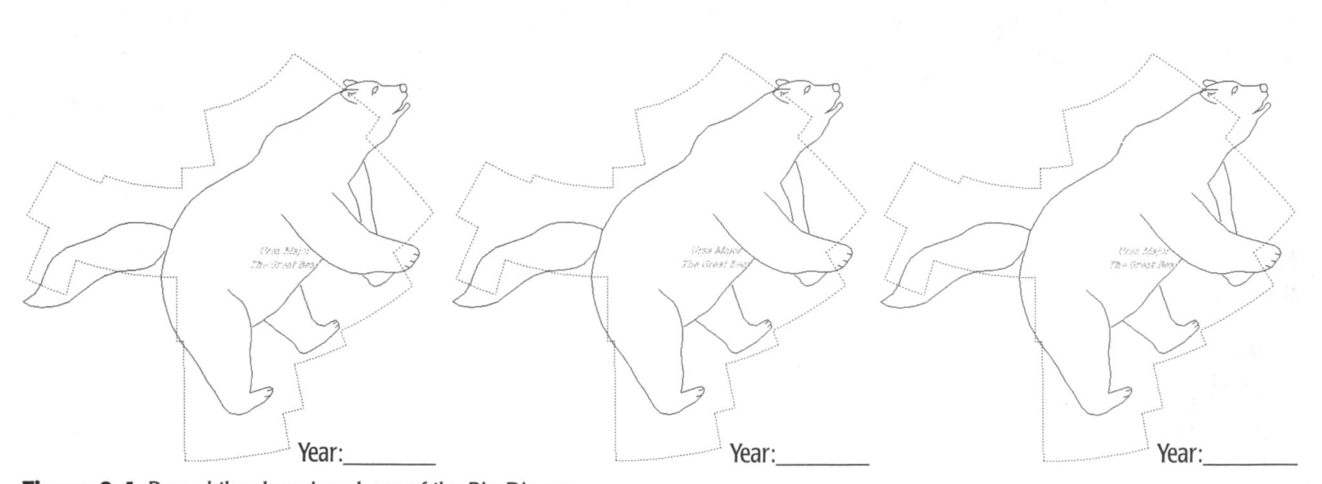

Figure 9-4 Record the changing shape of the Big Dipper

On Figure 9-5 below, follow the same procedure for Aquila the Eagle.

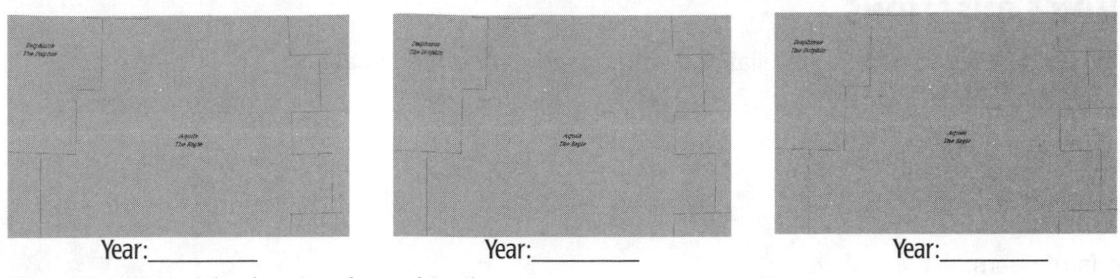

Figure 9-5 Record the changing shape of Aquila

Which constellation in the setting "Prehistoric Constellations" changed the most? What do you think is the reason for this?

CONCLUSION

In the space below, write a conclusion for this activity. Briefly explain what you did and what you learned from it.

CHECK YOUR UNDERSTANDING 9: PROPER MOTION

MULTIPLE-CHOICE QUESTIONS

1. The effects of proper motion on constellations are noticed over a time scale of
   a. years.
   b. decades.
   c. hundreds of years.
   d. thousands of years.
2. Which is an effect of proper motion?
   a. changing the pole star
   b. a change in the positions of the equinoxes
   c. changes in the relative positions of stars in constellations
   d. [all of the above ARE caused by proper motion]
3. Which will eventually cause the Big Dipper to "break-up"?
   a. precession
   b. proper motion
   c. [both will affect the Big Dipper in this way]
   d. [neither—unlike the Beatles, the Big Dipper will never "break-up"]
4. Proper motion is best observed for ______________ stars __________.
   a. close
   b. far
   c. bright
   d. [more than one of the above]
5. The effects of proper motion on nearby stars can be noticed over a time scale of
   a. days.
   b. months.
   c. years.
   d. [all of the above, depending on the distance to the star].
6. How much different would the constellations on a star chart drawn 2000 years ago differ from one drawn today?
   a. They would look much different.
   b. They would look a little different.
   c. They would look about the same.
   d. Star charts never change.

7. Stars that show observable proper motion over relatively short periods of time must be
   a. bright.
   b. dim.
   c. close.
   d. [both a and c]
8. Why do stars that show proper motion over relatively short periods of time appear to be dim?
   a. They are far away.
   b. The stars that are close enough just happen to be dim.
   c. No stars show proper motion over short time-scales.
   d. Actually, many bright stars show proper motion over short time-scales.
9. How do we know proper motion occurs?
   a. We see constellations changing.
   b. Ancient constellations were different than those we see today.
   c. Close stars can be observed to change position relative to father ones.
   d. [both a and b].
10. Proper motion is caused by.
   a. actual motion of stars through space.
   b. a combination of several motions of the Earth.
   c. precession.
   d. [none of the above].

OPEN-ENDED ACTIVITY

Use the skills you learned in this activity to observe the constellation Canis-Major and advance time far enough into the future and go back far enough into the past until it looks different. Record what Canis-Major looked like in the past and will look like in the future. Which star undergoes the most proper motion? Can you figure out why?

10
Phases of the Moon

INTRODUCTION

The Moon receives light from the Sun and as it orbits Earth the amount of area facing Earth that is illuminated by the Sun changes throughout the month. This results in differing amounts of the Moon disk being visible from Earth. These changes are known as the Moon's phases. See *MasteringAstronomy Tutorial* 4 for more about the Moon's phases.

PART 1: THE VIEW FROM SPACE

1. Select **Phases of the Planets** under the **Explore** menu. Then select **Phases of the Moon** (Figure 10-1).
2. Advance the time manually to each position on the chart on the RESULTS sheet. Use the display to color in the darkened portion of the Moon and label it with the correct phase name for each position. You will determine the rise and set times in Part 2.

Figure 10-1 Observe the phases of the Moon from above and from Earth

PART 2: RISING AND SETTING TIMES

File | Open Settings | Basics | Phase of the Moon

1. Under the **View** menu, use the command **Center Planet** and select **Moon** to bring up the Moon's **Data Panel.**
2. Change the time step on the **Time Panel** to **1 day** and manually click time backward until the Moon is 0 days old. The Moon's age is under the **General** tab on the **Data Panel.**
3. Advance the time to each of the phases observed in Part 1. Find the Moon's rising and setting times under the **Visibility** tab on the **Data Panel.** Record the times in the table on the RESULTS sheet.

PART 3: TYPES OF MONTHS

File | Open Settings | Basics | Moon Along the Ecliptic

1. Make sure that **Star Names** under the **Chart** menu is on. Zoom out or press the down arrow key until the star *Menkar* is visible in the frame.
2. Click on the Moon to bring up its **Data Panel** and record the information requested on the RESULTS sheet.
3. Change the time step on the **Time Panel** to **1 day** and advance the time manually in 1-day steps until *Menkar* is in the frame with the Moon again. It does not have to be in exactly the same place, merely in the frame (see Figure 10-2). Record the requested information in the RESULTS section again. Is the Moon in exactly the same phase and the same age as it was on the original date?
4. Since the Moon has returned to the same position relative to a star, it has completed one *sidereal month*. This is the time required for the Moon to orbit the Earth. Use your results to determine the approximate length of a sidereal month.
5. Now advance the time until the Moon is in exactly the same phase and the same age as it was on the original date. Record this date on the RESULTS sheet. Is *Menkar* still in the frame? If not, which star is in the frame?
6. The amount of time for the Moon to complete its phases is called a *synodic month*. Use your results to determine the approximate length of a synodic month.

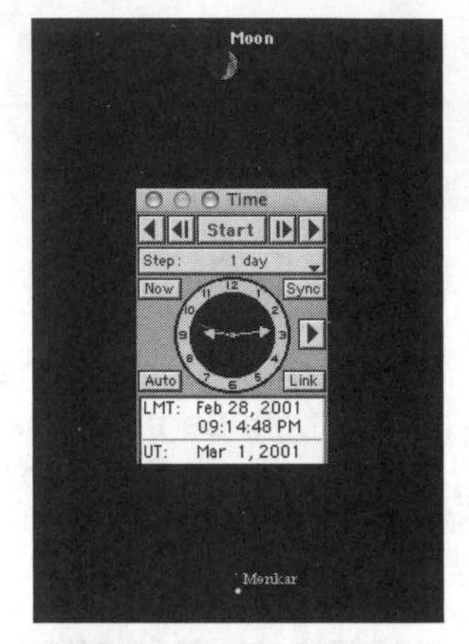

Figure 10-2 The **Time Panel** at the time that the Moon is near Menkar

PART 4: THE MONTHS FROM SPACE

File | Open Settings | Demos | Orbit of the Moon 2

1. Change the time step on the **Time Panel** to **1 day.**
2. Change the *Ecliptic Latitude* on the **Location Panel** to 90°.
3. Click on the image of the Moon on your Sky Chart to bring up its **Data Panel**.
4. Manually move the time backward one day so the Moon will have an age of 0 days. Your screen should look like Figure 10-3.
5. Advance the time manually until the Moon is 27 days old, the end of a sidereal month. Ignore the red and green drop lines meant to show the Moon's angle with the ecliptic. You can temporarily remove them by clicking the **eraser** icon in the last column on the **Planet Panel**.
6. Record the Moon's position in its orbit on the picture on the RESULTS sheet. Note how the Moon's position in its orbit compares to its original position.
7. Now advance the time until new Moon, the end of the synodic month, and record its position on the picture on the RESULTS sheet. Can you figure out the reason for the difference between the two types

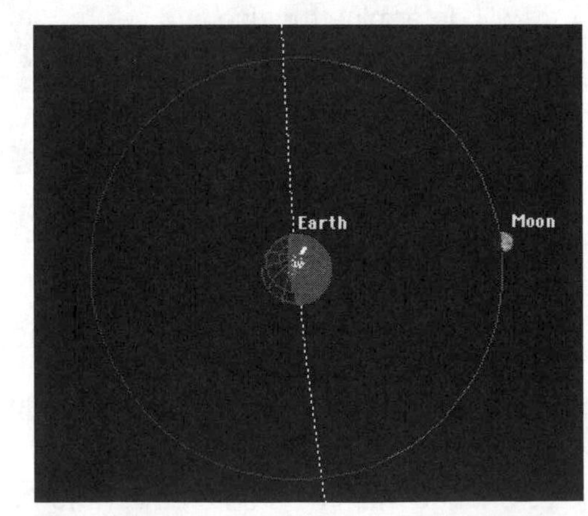

Figure 10-3 The Moon in its orbit during the new phase

of months? *Hint:* Look at the position of illuminated portion of Earth in the pictures on the RESULTS sheet.

PART 5: THE PHASES OF THE EARTH

1. Select **Phases of the Planets** under the **Explore** menu. Then select **Phases of the Earth** to observe Earth's phases from the Moon. Are they any different from the phases of the Moon observed from Earth? Answer the questions on the RESULTS sheet.

RESULTS SHEET 10 Phases of the Moon

NAME ______________________ **DATE** ______________ **SECTION** ______________________

PARTS 1 AND 2: THE VIEW FROM SPACE AND RISING AND SETTING TIMES

Use Figure 10-4 as a reference to fill in each circle with the correct phase. Then name the phase and record the Moon's rising and setting times for each numbered position in the figure.

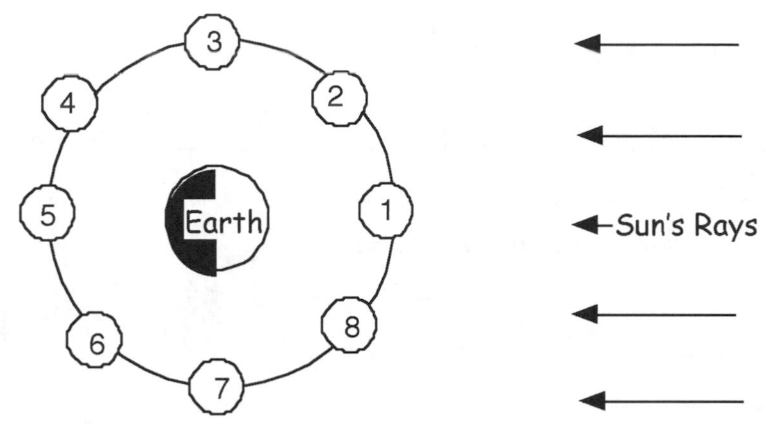

Figure 10-4 Lunar Phases diagram

	Phase Name	Rise/Set Times
1 ◯	______________	______________
2 ◯	______________	______________
3 ◯	______________	______________
4 ◯	______________	______________
5 ◯	______________	______________
6 ◯	______________	______________
7 ◯	______________	______________
8 ◯	______________	______________

In which phase will the Moon be up most of the night?

In which phase will the Moon be up most of the day?

In which phase is the Moon up about half the night, then half the day?

In which phase is the Moon up about half the day, then half the night?

PART 3: TYPES OF MONTHS

Original Date: __________ Moon Phase: __________ Age: ______

Date of next time *Menkar*

is in frame: __________ Moon Phase: __________ Age: ______

Length of sidereal month: ______ days

Phase and age that are the same as the original date:

Date: __________

Moon Phase: __________

Age: __________

Star in Frame: __________

Length of synodic month: ______ days

PART 4: THE MONTHS FROM SPACE

On Figure 10-5, record the position of the Moon in its orbit at ages 27 and 29 days.

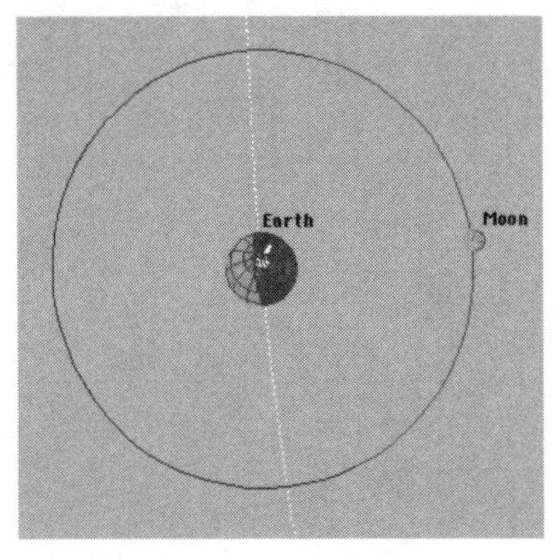

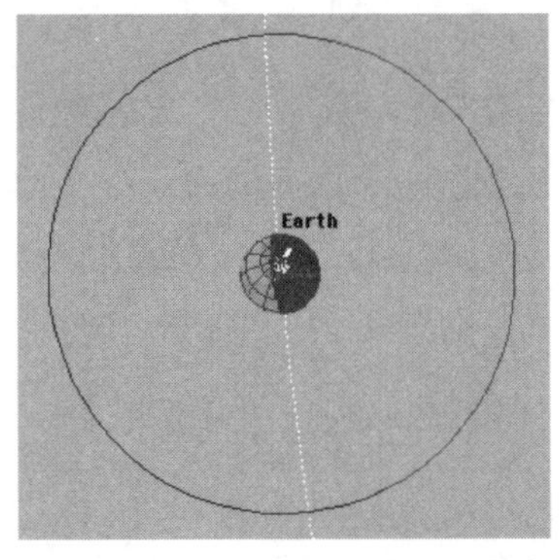

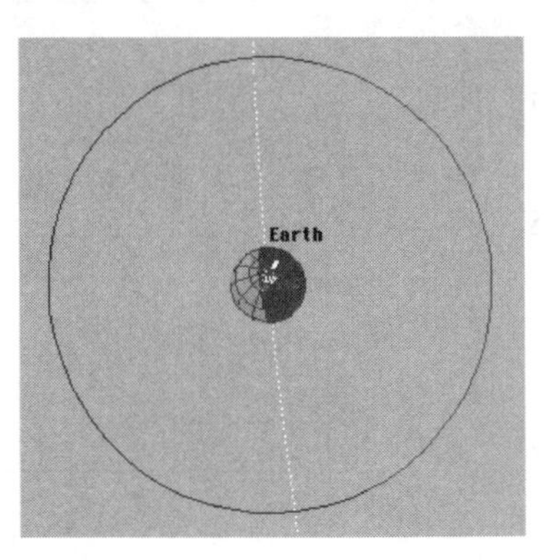

Figure 10–5 Record the positions of the moon at age 27 days and age 29 days

From looking at the three pictures in Figure 10-5, can you figure out why the synodic month is about two days longer than the sidereal month?

PART 5: THE PHASES OF THE EARTH

Does Earth seem to go through the same cycle of phases observed from the Moon as the Moon does when observed from Earth? **YES | NO** *(Circle one)*

As you observe Earth's phases, look at the positions of the Moon in its orbit and think about what phases of the Moon would be seen from Earth at the same times. Are they the same as Earth's phases seen from the Moon or different? Explain your answer.

Keeping in mind that the same side of the Moon always faces Earth, do you think that Earth would cycle through rising and setting times when viewed from the Moon in a similar way as the Moon does when viewed from Earth? Explain your answer.

CONCLUSION

In the space below, write a conclusion for this activity. Briefly explain what you did and what you learned from it.

CHECK YOUR UNDERSTANDING 10: PHASES OF THE MOON

MULTIPLE-CHOICE QUESTIONS

1. When Earth is in between the Sun and the Moon, the Moon is
   a. new.
   b. full.
   c. first quarter.
   d. last quarter.
2. When Earth is in between the Sun and the Moon, the Moon is
   a. up all day.
   b. up all night.
   c. up half the day, then half the night.
   d. up half the night, then half the day.
3. When the Moon is in between the Sun and the Moon, Earth is
   a. new.
   b. full.
   c. first quarter.
   d. last quarter.
4. When the Moon is in between the Sun and Earth, the Moon is
   a. up all day.
   b. up all night.
   c. up half the day, then half the night.
   d. up half the night, then half the day.
5. In which phase is the Moon up half the night, then half the day?
   a. full
   b. new
   c. first quarter
   d. last quarter
6. A first-quarter Moon is seen highest in the sky at
   a. noon.
   b. midnight.
   c. sunset.
   d. sunrise.

7. Which month is longer?

   a. synodic

   b. sidereal

   c. [they are the same length]

8. In which month is a Blue Moon (two full Moons in one calendar month) impossible?

   a. October

   b. February

   c. May

   d. [this is never possible]

9. Which month is observable from Earth?

   a. synodic

   b. sidereal

   c. [both are observable from Earth]

10. When a full Moon is observed from Earth, an observer on the Moon would see a ________________ Earth.

   a. full

   b. new

   c. quarter

   d. crescent

OPEN-ENDED ACTIVITY

Using the skills you learned in this activity, determine the dates of the next new, first-quarter, full, and last-quarter moons visible from your location.

11
Solar Eclipses

INTRODUCTIONS

A *solar eclipse* occurs when the Moon, in its orbit around Earth, crosses the line between Earth and the Sun. This does not happen every month. When it does happen, only observers at the locations on which the Moon casts its shadow will see the eclipse. In this activity you will learn the reasons for these conditions while observing several of these rare and exciting events. See *MasteringAstronomy Tutorial* 3 for more about eclipses.

PART 1: SOLAR ECLIPSE TYPES

1. Select **Shadows on the Earth** under the **Explore** menu.
2. Advance the time manually and observe the **View from the Earth** for a **Total Eclipse**, an **Annular Eclipse**, a **Partial Eclipse**, and **No Eclipse**. Record the position on the RESULTS sheet of the Moon during each eclipse in the space provided.

PART 2: WHEN DO THEY OCCUR?

File | Open Settings | Basics | Moon along the Ecliptic

1. Click on the Moon to open its **Data Panel** (Figure 11-1).
2. Change the time step to **1 day** and manually advance until the next new Moon (0 days old).
3. Now change the time step to **synodic month** as is shown in Figure 11-2. Manually advance the time one month at a time until you find a month where the Moon is on (or very near) the ecliptic.
4. When you find a month where the Moon is on or near the ecliptic, change the time step to **10 minutes** and move time backward and/or forward to see if the Moon moves in front of the Sun at any time. If it does, you have found a solar eclipse. Record on the RESULTS sheet the date of the first solar eclipse you find.

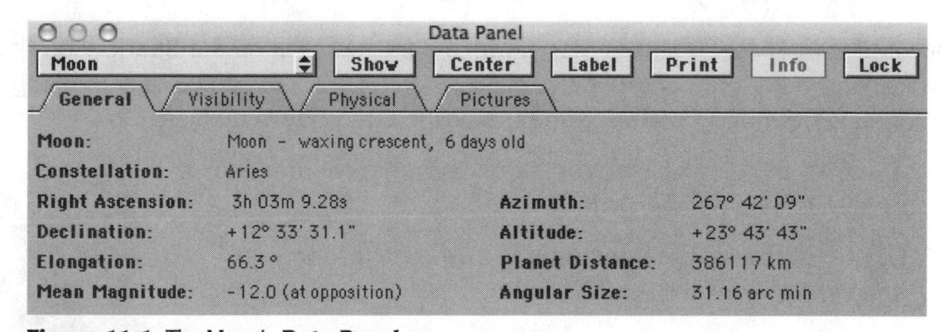

Figure 11-1 The Moon's **Data Panel**

5. Zoom in until the Moon and the Sun fill your Sky Chart to get a good look at the eclipse. Does it appear to be total or partial?
6. Move the **location marker** ⊕ on the **Location Panel** to your location. Does the eclipse appear to be visible from your location? Adjust time forward and backward in 10-minute steps to observe the entire eclipse to be sure. Make note of whether the Moon covers the Sun and if the Sun is visible from your location during the time of the eclipse.
7. Move the **location marker** ⊕ on the **Location Panel** to different positions until you find a place where the eclipse appears to be total, or very close to total. You may have to switch to a smaller time step to be sure of this. Record this location on the RESULTS sheet.

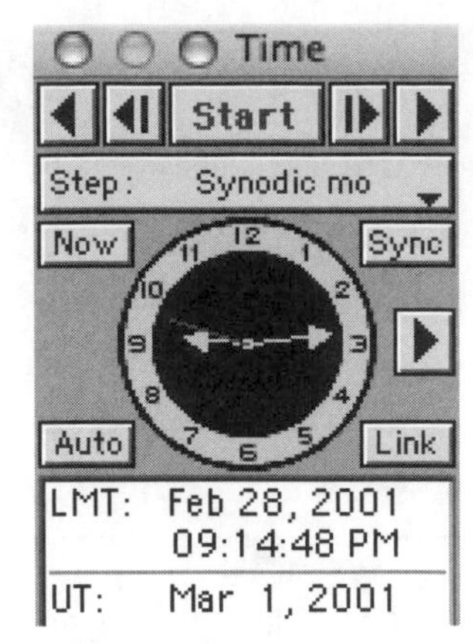

Figure 11-2 The **Time Panel** set for a time step of one synodic month

PART 3: A FUTURE ECLIPSE

File | Open Settings | Demos | America Eclipse 2017—two views

1. Change the time step to **1 minute**.
2. Go backward in time to before the start of the eclipse (see Figure 11-3). Then advance the time and watch the eclipse. Record the information requested on the RESULTS sheet.
3. Return to a time before the eclipse, select **Earth** on the **Location Panel**, and move the **location marker** ⊕ to your home location.
4. Advance the time forward and watch the eclipse again, this time from home. Record the requested information on the RESULTS sheet.

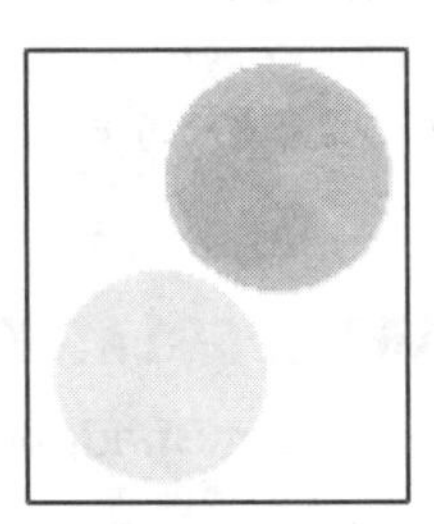

Figure 11-3 The Sun and the new Moon just before a solar eclipse

PART 4: THE ECLIPSE PATH

File | Open Settings | Basics | Shadow over America

This setting is a view from the Moon. Make sure you click on the **Shadow** on the **Planet Panel.** The inner (umbra) and outer (penumbra) parts of the Moon's shadow can be seen moving across the Earth over the locations that will be able to see the eclipse, as shown in Figure 11-4.

1. Advance the time in 1-minute steps and observe the Moon's shadow move across Earth.
2. Draw the eclipse path (the path followed by the *inner* shadow) on your RESULTS sheet.
3. Now select **Earth** on the **Location Panel** and move your **location marker** ⊕ to somewhere that was along the path of the inner shadow.
4. Use the **Planet Panel** to *center* (click on box to highlight in red) and *lock* (click on **lock** icon) the Sun.
5. Use the time step on the **Time Panel** to watch the eclipse.
6. Do the same for a location that was in the path of the outer shadow and a location that was not in the shadow at all. Fill out the table and answer the questions on the RESULTS sheet.

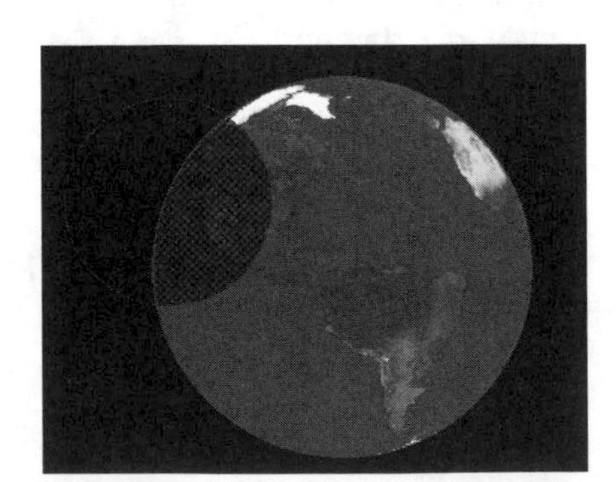

Figure 11-4 The Moon's Shadow cast on Earth

PART 5: PAST ECLIPSES

Observe some other solar eclipses:

File | Open Settings | Demos | African Eclipse-June 2001
File | Open Settings | Settings | Eclipse 2001 from Zambia
File | Open Settings | Basics | Christmas Eclipse 2000
File | Open Settings | Demos | Europe Eclipse 1999
File | Open Settings | Settings | Eclipse 1991
File | Open Settings | Settings | Eclipse 1991 from Moon
File | Open Settings | Demos | Eclipse in China 709 BC
File | Open Settings | Demos | Greek Eclipse 413 BC

RESULTS SHEET 11 Solar Eclipses

NAME ______________________ **DATE** ______________ **SECTION** ______________________

PART 1: SOLAR ECLIPSE TYPES

Record the position of the Moon in each case in the space provided.

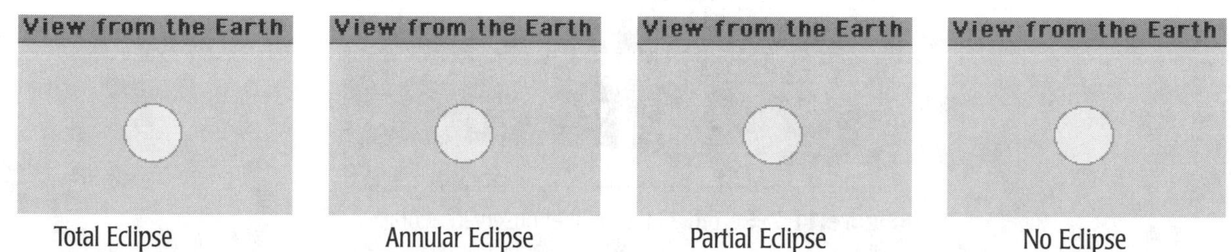

Figure 11-5 Record the position of the Moon for each type of eclipse

How does the apparent size of the Moon during an annular eclipse compare to its apparent size during a total eclipse?

What could cause the apparent difference?

What is the phase of the Moon during a solar eclipse? *Hint:* Looking at the **View from Above the Earth and Moon** may help.

PART 2: WHEN DO THEY OCCUR?

Date of first solar eclipse found: ______________

Was the eclipse total? ______________

On Figure 11-6, draw the Moon covering the amount of the Sun you observed from your location.

Was the eclipse visible from your location?

If it was visible, was it TOTAL | PARTIAL? *(circle one)*

Sun

Figure 11-6 Draw the Moon covering the amount of the Sun you observed

Record on the figure below the approximate position that observed a total eclipse. The **location marker** shown is on San Francisco, which is the default location of *SkyGazer.*

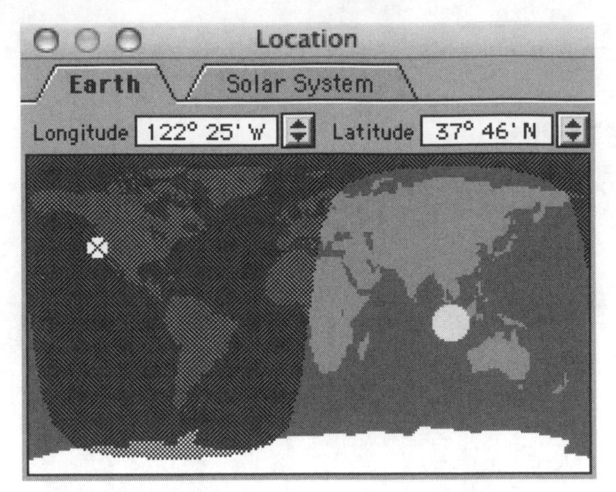

Figure 11-7 Record a location that observed a total eclipse

Why wasn't there a solar eclipse every month? *Hint:* Think about where the Moon was when there *was* an eclipse.

PART 3: A FUTURE ECLIPSE

Record the information about the eclipse on the table below.

	Casper, Wyoming	Your Location
Time Eclipse Began		
Time Totality Began		
Time Totality Ended		
Time Eclipse Ended		
Duration of Totality		
Duration of Eclipse		

(If the eclipse was *not* total from your location, you cannot record times and duration for totality.)

In the space below, draw the disk of the Moon over the Sun at the time of maximum coverage as seen from your location.

Time of maximum coverage: ____________________

Was the eclipse total as seen from your location?

Did it get dark at your location?

PART 4: THE ECLIPSE PATH

Draw the eclipse path (the path of the center of the shadow) on Earth, using Figure 11-8.

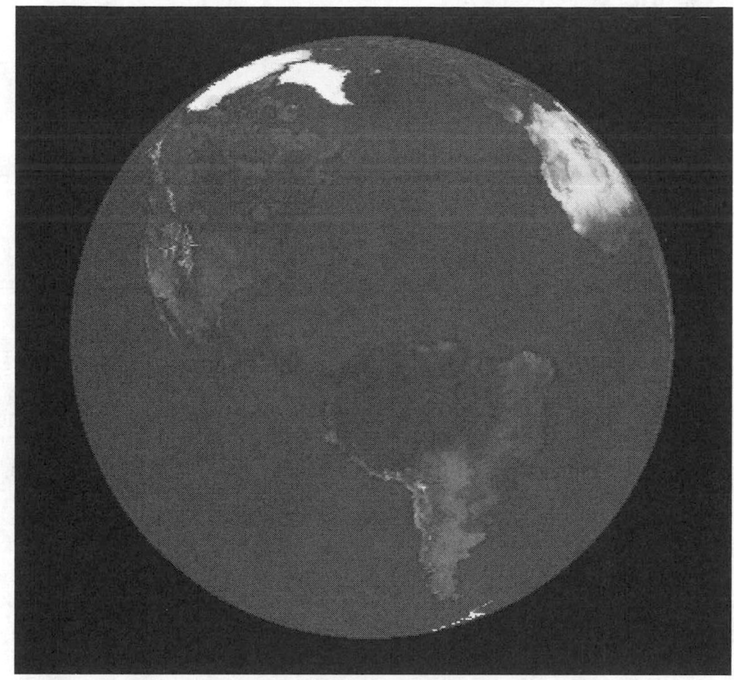

Figure 11-8 Draw the path of the center of the Moon's shadow over the surface of Earth

Record on Figure 11-9 the three locations from which you observed the eclipse. The location marked on the map is in South Carolina, where the eclipse will be total.

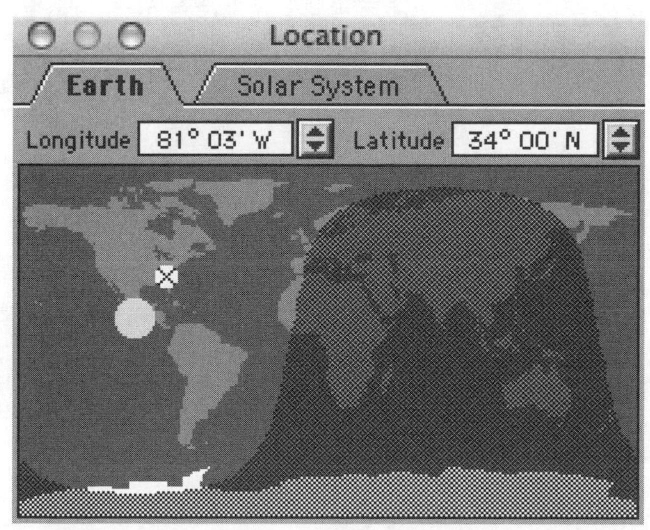

Figure 11-9 Record the locations from which you observed the eclipse

Using the table below, record the locations from which you saw a total eclipse, a partial eclipse, or no eclipse. You may record either latitude and longitude or a city name (or both).

	Total Eclipse	Partial Eclipse	No Eclipse
Location on the path of the inner shadow		X	X
Location on the path of the outer shadow	X		X
Location not on the path of either shadow	X	X	

To see a total eclipse, the INNER I OUTER *(circle one)* shadow must pass over the location.

To see a partial eclipse, the INNER I OUTER *(circle one)* shadow must pass over the location.

A location that is not in the shadow will see _________ eclipse.

CONCLUSION

In the space below, write a conclusion for this activity. Briefly explain what you did and what you learned from it.

CHECK YOUR UNDERSTANDING 11: SOLAR ECLIPSES

MULTIPLE-CHOICE QUESTIONS

1. A solar eclipse will occur
   a. at every full Moon.
   b. during every new Moon.
   c. at full Moon only when the Moon is on the ecliptic.
   d. at new Moon only when the Moon is on the ecliptic.
2. During a solar eclipse,
   a. the Moon is in between Earth and the Sun.
   b. the Sun is in between Earth and the Moon.
   c. the Earth is in between the Sun and the Moon.
   e. [none of the above are true].
3. A solar eclipse occurs when
   a. the Moon covers up the Sun.
   b. the Sun covers up the Moon.
   c. Earth's shadow covers the Moon.
   d. the Moon's shadow covers Earth.
4. During a total solar eclipse,
   a. the Moon turns reddish in color.
   b. you can see the outer layers of the Sun around the disk of the Moon.
   c. it gets dark in the middle of the day.
   d. [more than one of the above occurs].
5. Totality during a solar eclipse can last
   a. only a few seconds.
   b. a few minutes.
   c. up to about an hour.
   d. [it can vary from eclipse to eclipse].
6. An entire solar eclipse lasts
   a. only a few minutes.
   b. about an hour.
   c. a few hours.
   d. [it can vary from eclipse to eclipse].

7. If an observer on Earth sees a total solar eclipse,
   a. everyone on the daytime side of Earth is seeing it.
   b. someone elsewhere on Earth must be seeing an annular eclipse.
   c. someone elsewhere on Earth must be seeing a partial solar eclipse.
   e. someone elsewhere on Earth must be seeing a lunar eclipse.
8. If an observer on Earth sees a partial solar eclipse,
   a. everyone on the daytime side of the Earth is seeing it.
   b. someone elsewhere on Earth must be seeing an annular eclipse.
   c. someone elsewhere on Earth must be seeing a total solar eclipse.
   d. [none of the above has to be true].
9. Annular eclipses occur
   a. when the Moon is closer to Earth in its orbit.
   b. when the Moon is farther from Earth in its orbit.
   c. when the disk of the Moon appears to be its largest.
   d. [more than one of the above].
10. If a solar eclipse is observed from Earth, an observer on the Moon would see
    a. a lunar eclipse.
    b. a solar eclipse.
    c the Moon's shadow covering the whole Earth (a terran eclipse).
    d. a spot on Earth covered by the Moon's shadow.

OPEN-ENDED ACTIVITY

Using the skills you learned in this activity, try to determine the date of the most recent solar eclipse visible from your location and whether it was full or partial. Do the same for the next solar eclipse visible from your location.

12

Lunar Eclipses

INTRODUCTION

A *lunar eclipse* occurs when Earth moves directly between the Moon and the Sun and casts its shadow onto the Moon. In this activity you will learn about the conditions necessary for lunar eclipses and observe several of them. See *MasteringAstronomy Tutorial* 3 for more about eclipses.

PART 1: VIEWING A LUNAR ECLIPSE

File I Open Settings I Basics I Lunar Eclipse

1. Open the **Location Panel** and move the **location marker ¢** to your location.
2. Step back in time to the beginning of the eclipse when the Moon first enters the inner, darker part of the shadow, called the *umbra*. The outer part of Earth's shadow is *penumbra*.
3. Advance the time and watch the entire eclipse. Record the information requested on the RESULTS sheet.
4. Now return to a time during totality. Use the **Location Panel** to try other locations on Earth and see if the eclipse is visible. Remember, if the Moon is below the horizon, you cannot see it. Answer the question on the RESULTS sheet.

PART 2: WHEN WILL THEY OCCUR?

File I Open Settings I Demos I Moon Orbit

1. Click on the **Zoom** button and select 10°.
2. Bring up the **Planet Panel** under the **Control** menu and deselect both the **Path** and **Trail** for the Moon.
3. Under the **Display** menu select **Reference Markers**. Select **Ecliptic Equator** and make sure all the others are deselected.
4. Advance the time automatically in 1-hour steps until you notice that the Moon appears full. Now click on the Moon to bring up its **Data Panel** and click on the **Visibility** to advance time a few more hours watching the **Illumination** until it reaches 100%. Now change time-step to **synodic month** and advance time until you notice a lunar eclipse. Record the required information on the RESULTS sheet.
5. Now, starting from the time of the lunar eclipse you just found, with the time step still **synodic month** and advance the time manually until you find another lunar eclipse.
6. Once you find a lunar eclipse, change the time step to **1 minute** and adjust the time back and forth to observe the eclipse. Answer the questions on the RESULTS sheet.

PART 3: TURNABOUT

File | Open Settings | Basics | Shadow over America

1. This setting is a view from the Moon. Use **Set Time** under the **Control** menu to change to the date and time of the first lunar eclipse you observed in Part 1.
2. Use the **Planet Panel** under the **Control** menu to center and lock Earth.
3. Advance the time in 10-minute steps and see what happens. Report your observations on the RESULTS sheet.

PART 4: GEOMETRY OF ECLIPSES

File | Open Settings | Demos | Orbit of the Moon 2

1. Go to the **Solar System** tab of the **Location Panel** (pictured in Figure 12-1) to change **Latitude** to 90°. This allows you to view the Earth-Moon system from above.
2. Use **Set Time** under the **Control** menu to change to the date of the first lunar eclipse you observed in Part 1.
3. Use the **Planet Panel** under the **Control** menu to center and lock Earth. Record the requested information on the RESULTS sheet.

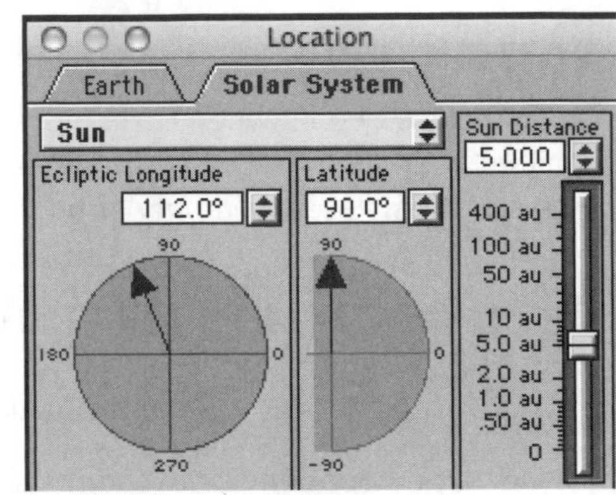

Figure 12-1 The **Location Panel** set for 90° Latitude

PART 5: A FAMOUS ECLIPSE IN HISTORY

File | Open Settings | Settings | Columbus' Eclipse of 1504

1. Advance the time to observe Columbus' Eclipse and answer the questions on the RESULTS sheet.
2. Use **Set Location** under the **Control** menu to determine the location from which he observed the eclipse.

RESULTS SHEET 12 Lunar Eclipses

NAME ______________________ **DATE** ____________ **SECTION** ____________

PART 1: VIEWING A LUNAR ECLIPSE

Date of the lunar eclipse: __________

Record the information about the eclipse on the table below.

	Your Location
Time eclipse began (Moon entered umbra)	
Time totality began (Moon completely covered by umbra)	
Time totality ended (Moon no longer completely in umbra)	
Time eclipse ended (Moon left umbra)	
Duration of totality	
Duration of eclipse	

In general, where on the Earth was the eclipse visible and where was it not visible?

PART 2: WHEN WILL THEY OCCUR?

Date of first lunar eclipse: __________

Eclipse was UMBRAL I PENUMBRAL *(circle one)*.

Date of next lunar eclipse: __________

Eclipse was UMBRAL I PENUMBRAL *(circle one)*.

Does a lunar eclipse occur every month?

For a lunar eclipse to occur, the ______________________ (phase) Moon must be on or very near the ______________________ (reference line). The intersection points between the Moon's orbit and this line are called ______________________ (if you don't know this answer, make sure to read the **Settings File Banner** (the blue box)).

PART 3: TURNABOUT

When a lunar eclipse is seen from the Earth, a ______________ ______________ is seen from the Moon.

PART 4: GEOMETRY OF ECLIPSES

On Figure 12-2, draw and label the Moon in its position in its orbit during a lunar eclipse. Draw a line from the Moon through Earth in the direction of the Sun. This would be the line of sight that would allow someone on the Moon to see what you observed in Part 3.

Using Figure 12-2 again, draw and label the Moon's position during a solar eclipse.

To see a lunar eclipse you must be on the DAYTIME I NIGHTTIME *(circle one)* side of Earth.

To see a solar eclipse you must be on the DAYTIME I NIGHTTIME *(circle one)* side of Earth.

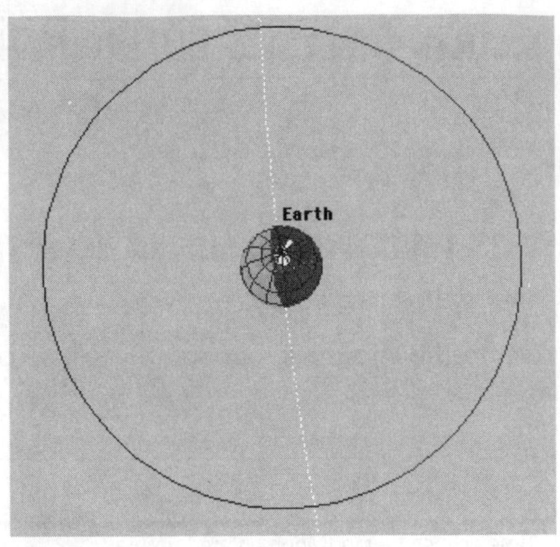

Figure 12-2 Draw the Moon's position on its orbit during a lunar eclipse

PART 5: A FAMOUS HISTORICAL ECLIPSE

Make sure to read the **Settings File Banner**.

Why is the 1504 eclipse called Columbus'?

What kind of eclipse was it?

Was it total?

About what time of day did it occur?

How did he know about it?

Near what modern-day city did he observe the eclipse?

CONCLUSION

In the space below, write a conclusion for this activity. Briefly explain what you did and what you learned from it.

CHECK YOUR UNDERSTANDING 12: LUNAR ECLIPSES

MULTIPLE-CHOICE QUESTIONS

1. A lunar eclipse will occur
   a. at every full Moon.
   b. during every new Moon.
   c. at full Moon only when the Moon is on the ecliptic.
   d. at new Moon only when the Moon is on the ecliptic.
2. During a lunar eclipse,
   a. the Moon is in between Earth and the Sun.
   b. the Sun is in between Earth and the Moon.
   c. Earth is in between the Sun and the Moon.
   d. [none of the above are true].
3. A lunar eclipse occurs when
   a. the Moon covers up the Sun.
   b. the Sun covers up the Moon.
   c. Earth's shadow covers the Moon.
   d. the Moon's shadow covers Earth.
4. During a total lunar eclipse,
   a. the Moon turns reddish in color.
   b. you can see the outer layers of the Sun around the disk of the Moon.
   c. it gets dark in the middle of the day.
   d. [more than one of the above occurs].
5. Totality during a lunar eclipse can last
   a. only a few seconds.
   b. a few minutes.
   c. up to about an hour.
   d. it can vary from eclipse to eclipse.
6. An entire lunar eclipse lasts
   a. only a few minutes.
   b. about an hour.
   c. a few hours.
   d. it can vary from eclipse to eclipse.

7. If an observer on Earth sees a total lunar eclipse,
   a. everyone on the nighttime side of Earth is seeing it.
   b. someone elsewhere on Earth must be seeing a total solar eclipse.
   c. someone elsewhere on Earth must be seeing a partial lunar eclipse.
   d. [none of the above are ever true].
8. If an observer on Earth sees a partial lunar eclipse,
   a. everyone on the nighttime side of Earth is seeing it.
   b. someone elsewhere on Earth must be seeing a partial solar eclipse.
   c. someone elsewhere on Earth must be seeing a total lunar eclipse.
   d. [none of the above are ever true].
9. You are more likely to see a __________ eclipse in your lifetime, because they are
   a. solar; more common than lunar eclipses.
   b. solar; seen over a greater area of Earth's surface than lunar eclipses.
   c. lunar; more common than solar eclipses.
   d. lunar; seen over a greater area of Earth's surface than solar eclipses.
10. If a total lunar eclipse is observed from Earth, an observer on the Moon would see
    a. a total lunar eclipse.
    b. a total solar eclipse.
    c. a total eclipse of Earth.
    d. [possibly more than one of the above, depending on the geometry of the eclipse].

OPEN-ENDED ACTIVITY

Using the skills you learned in this activity, try to determine the date of the most recent lunar eclipse visible from your location and whether it was full or partial. Do the same for the next lunar eclipse visible from your location.

VOYAGER: SKYGAZER CD-ROM ACTIVITIES

13

The Inferior Planets

INTRODUCTION

The *inferior planets* are those planets that are closer to the Sun than Earth. Their motion will be the focus of this activity. The planets farther from the Sun than Earth, the *superior planets*, will be the focus of the next activity.

PART 1: WHICH PLANETS ARE THE INFERIOR PLANETS?

1. Select **The Solar System** under the **Explore** menu.
2. Under the **Show** section of **The Solar System** display, deselect the **Outer Planets**.
3. Then use the **View Location** scroll bar to change from a distance of 50 AU to a distance of about 3 AU to see which planets are closer to the Sun than Earth. Record the names of these inferior planets on the RESULTS sheet.
4. Advance the time and watch the planets orbit the Sun. Answer the questions on the RESULTS sheet.

PART 2: THE PERIODS OF VENUS

File | Open Settings | Basics | Planet Orrery

1. Adjust the **Sun Distance** on the **Location Panel** to about 5 AU.
2. Use the **Planet Panel** under the **Control** menu to turn Earth and its **Name** on. Your Sky Chart should look similar to the one pictured in Figure 13-1.
3. Set the time to today's date and record this on the RESULTS sheet.
4. Change your time step to **1 day** and advance the time until Venus is between Earth and the Sun. When an inferior planet is between Earth and the Sun, the configuration is called *inferior conjunction*. Record the date of Venus's inferior conjunction on the RESULTS sheet.
5. Now advance the time until the Sun is between Earth and Venus. This configuration is called *superior conjunction*. Record this date on your RESULTS sheet.
6. Advance the time until Venus is again at inferior conjunction. Record the date of this inferior conjunction and determine the approximate *synodic period* of Venus.

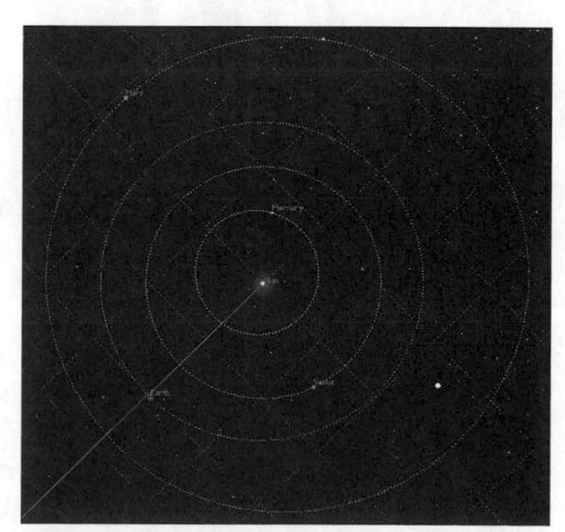

Figure 13-1 Mercury and Venus are the inferior planets

This is the amount of time between two identical configurations between the planet, Earth, and the Sun. Compare your value of Venus's synodic period to the value given under the **Physical** tab in Venus's **Data Panel**.

7. Take note of the position of Venus in its orbit around the Sun. Advance time again, but this time ignore Earth. Stop when Venus has returned to approximately the same position in its orbit around the Sun that it was when you started. One of the gridlines in your Sky Chart may be a useful starting and stopping point. The amount of time required for a planet to complete one orbit around the Sun is called its *sidereal period*. Record Venus's approximate sidereal period. Compare your value of Venus's sidereal period to the value given under the **Physical** tab in Venus's **Data Panel**.

PART 3: COMPARING THE PERIODS

1. You can observe the difference between an inferior planet's synodic period and sidereal period by selecting **Paths of the Planets** under the **Explore** menu. Select **Venus Chases the Sun**.
2. Advance time until Venus is at inferior conjunction. Use the **View from Earth** window, to help you (see Figure 13-2).
3. Now advance the time until Venus returns to approximately the same position in its orbit around the Sun (you can use whatever constellation Venus is in as a guide).

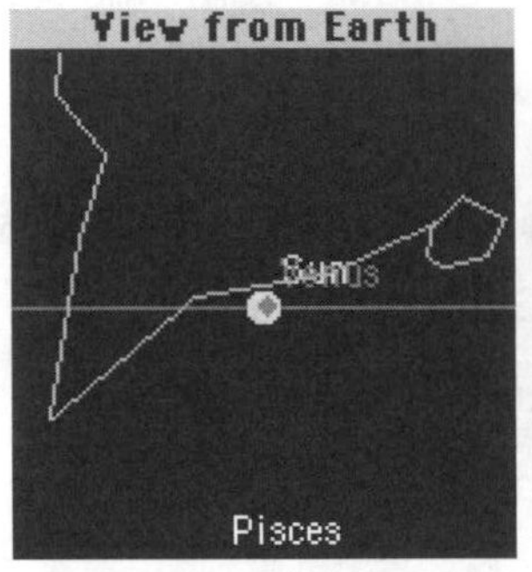

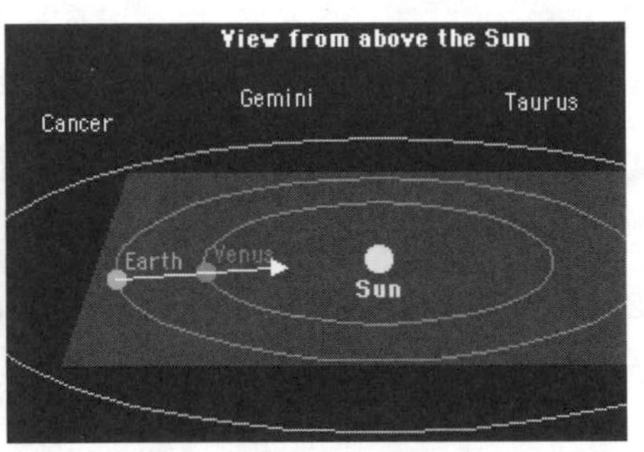

Figure 13-2 A view of Venus from Earth and its orbit from above the Sun

4. Is Venus at inferior conjunction again? If not, continue to advance the time until the next inferior conjunction. Answer the questions on the RESULTS sheet.

PART 4: SEEING VENUS FROM EARTH

1. Select **Phases of the Planets** under the **Explore** menu.
2. Select **Phases of Venus** and advance the time until Venus is at inferior conjunction.
3. Record the phase of Venus and draw it in the box provided on the RESULTS sheet.
4. Now advance the time and observe the phases until Venus is at superior conjunction.
5. Record the phase at superior conjunction on the RESULTS sheet.
6. Pay special attention to the phase and size of Venus during the two conjunctions and answer the questions on the RESULTS sheet.
7. Now return to your Sky Chart window. On the **Location Panel**, change the viewpoint from **Solar System** to **Earth**. Select **Center Planet** under the **View** menu and select **Venus**. Venus's **Data Panel** should come up. If it does not, click on Venus's symbol and it will. Make sure that Venus is still at superior conjunction (it should be lined up with the Sun). If it is not, adjust the time until it is.

Planet Panel

Planets | Comets | Asteroids | Satellites

Planets	Name	Sym	Orbit	Path	Trail
Sun	X				
Mercury	X	X	X		
Venus	X	X	X		
Earth		X	X		
Mars	X	X	X		
Jupiter	X	X	X		
Saturn	X	X	X		
Uranus	X	X	X		
Neptune	X	X	X		
Pluto	X	X	X		
Moon	X	X			
Shadow		X			

Figure 13-3 The **Planet Panel** with all planets but Venus deselected

From the **General** tab, record: Venus's **Magnitude, Angular size, Distance** (this is the distance from the Earth), and the **Sun Distance** (to be used in Part 5) in the table on the RESULTS sheet.

8. Use the **Planet Panel** to lock Venus and deselect the other planets to avoid confusion. Leave the Sun on (see Figure 13-3).
9. Now advance the time until Venus is at inferior conjunction (this should be the next time Venus is lined up with the Sun). Record the information from the **Data Panel** requested on the RESULTS sheet.
10. Answer the questions on the RESULTS sheet.

File | Open Settings | Basics | Phases of Mercury

11. Observe Mercury's phases as seen from Venus and answer the questions on the RESULTS sheet.

PART 5: KEPLER'S LAWS

File | Open Settings | Basics | Planet Orrery

You can use this setting to verify Kepler's Laws of Planetary Motion. Mercury gives a good example of the first law, *The Law of Elliptical Orbits*.

1. Zoom in until the orbit of Venus is the outermost orbit in your Sky Chart.
2. Advance the time in 1-day steps and observe Mercury's orbit. Draw Mercury's orbit in Figure 13-5 on the RESULTS sheet.
3. Kepler's second law is *The Law of Equal Areas*. To verify this law, select **Other** for the time step and set it to 11.00 days.
4. Record Mercury's initial position on your drawing of its orbit on the RESULTS sheet.
5. Manually advance time in 11-day steps and record Mercury's position with a dot after each step on Figure 13-5 until it has returned to its original position. Record Mercury's sidereal period on the RESULTS sheet.
6. Now draw lines connecting the Sun to each dot in the orbit that you marked. Observe the areas of the wedges you have drawn and answer the questions on the RESULTS sheet.

Kepler's third law takes the form of an equation

$$P^2 = a^3$$

where a is the planet's average distance from the Sun in its elliptical orbit (which is also the semimajor axis of the ellipse) and P is the sidereal period of the orbit.

7. Test the third law for Venus's orbit. First you need the orbital period in years. Divide the sidereal period you determined in Part 2 (in days) by 365. Then square the period P (in years).
8. Venus's orbit is not very elliptical (it is of low eccentricity), so the distance from the Sun does not vary much. Round off the values you recorded for Venus's **Sun Distance** in Part 4 to two digits after the decimal point; you will find that they are very close, if not the same. Cube the distance a. Record your calculations on the RESULTS sheet.
9. Return to the Sky Chart window and try this for Mercury's orbit. As you may have noticed when you drew it, Mercury's orbit is much more elliptical than that of Venus or Earth (it is of much higher eccentricity). This means that Mercury's **Sun Distance** will vary throughout the orbit, and the average must be used in Kepler's third law.

10. Advance time in 1-day steps until Mercury appears to be approximately at perihelion—the point in its orbit closest to the Sun. Record Mercury's **Sun Distance** from the **Data Panel**. Label the position of perihelion on your drawing of Mercury's orbit on the RESULTS sheet.
11. Then advance the time to the farthest point from the Sun in Mercury's orbit, *aphelion*. Record the **Sun Distance** and mark the position on your drawing.
12. Test Kepler's third law for Mercury and record the results.

RESULTS SHEET 13 The Inferior Planets

NAME ______________________ **DATE** ______________ **SECTION** ______________________

PART 1: WHICH PLANETS ARE THE INFERIOR PLANETS?

Which planets are considered the inferior planets?

The inferior planets orbit the Sun CLOCKWISE I COUNTERCLOCKWISE *(circle one)*.

PART 2: THE PERIODS OF VENUS

Start Date: ______________

Date of Venus's inferior conjunction: ______________

Date of Venus's superior conjunction: ______________

Date of Venus's next inferior conjunction: ______________

Synodic period of Venus = ______________ days

Synodic period from **Data Panel** = ______________ days

Sidereal period of Venus = ______________ days

Sidereal period from **Data Panel** = ______________ years × 365 = ______________ days

In Figure 13-4, draw Venus in its orbit at the positions of both inferior and superior conjunction.

PART 3: COMPARING THE PERIODS

After starting at inferior conjunction and completing one sidereal period, had Venus returned to inferior conjunction? YES I NO *(circle one)*

The SYNODIC I SIDEREAL *(circle one)* period of Venus is longer.

Does this agree with your results from Part 2? YES I NO *(circle one)*

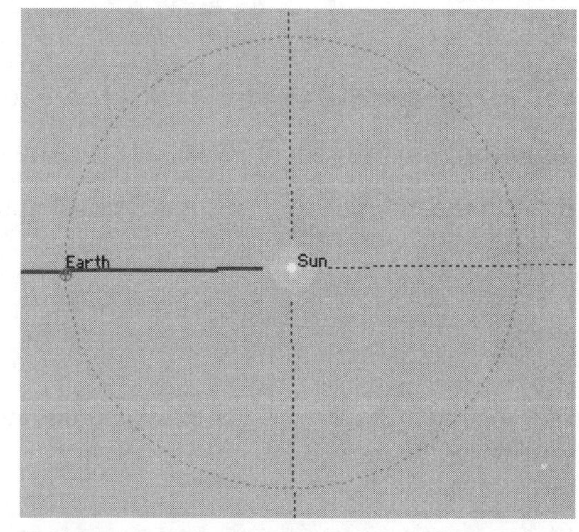

Figure 13-4 Draw Venus' position at both inferior and superior conjunction

PART 4: SEEING VENUS FROM EARTH

In each box below draw the phase of Venus that you observed.

Inferior Conjunction **Superior Conjunction**

Phase: ______________ Phase: ______________

What difference (other than phase) do you notice in the appearance of Venus in the two configurations above? What is the reason for this difference?

Fill out the information in the table below from Venus's **Data Panel**.

	Inferior Conjunction	**Superior Conjunction**
Magnitude		
Angular Size		
Distance (from Earth)		
Sun Distance		

Venus appears brightest at INFERIOR | SUPERIOR *(circle one)* conjunction.

Venus appears largest at INFERIOR | SUPERIOR *(circle one)* conjunction.

Explain the reasons for the changes in Venus's apparent brightness and size.

Did you ever observe Venus changing direction in its motion through the sky? This is known as *retrograde motion*.

Does Mercury go through the same cycle of phases observed from Venus as Venus goes through observed from Earth? How would this compare to Mercury's phases observed from Earth?

Sun

Figure 13-5 Draw Mercury's orbit

PART 5: KEPLER'S LAWS

Draw the orbit of Mercury on Figure 13-5.

What is the shape of Mercury's orbit?

Is the Sun at the center?

If the area of each wedge is equal, when is Mercury traveling the fastest in its orbit?

When is Mercury traveling the slowest?

Mercury's Perihelion Distance: ______________ AU

Aphelion Distance: ______________ AU

Mercury's average distance from the Sun: ______________ AU

Period of Mercury's orbit: ______________ days / 365 = ______________ years

	Period (days)	Period (yrs)	Sun Distance (AU)	P^2	a^3
Mercury					
Venus					

Does Kepler's third law seem to hold for Mercury and Venus?

CONCLUSION

In the space below, write a conclusion for this activity. Briefly explain what you did and what you learned from it.

CHECK YOUR UNDERSTANDING 13: THE INFERIOR PLANETS

MULTIPLE-CHOICE QUESTIONS

1. Which planet is not considered inferior?
   a. Mercury
   b. Venus
   c. Mars
   d. [all of the above ARE inferior]
2. In which phase will an inferior NOT be observed?
   a. new
   b. full
   c. gibbous
   d. [actually, inferior planets go through an entire cycle of phases]
3. Which period of an inferior planet is longer?
   a. synodic
   b. sidereal
   c. [both periods are the same]
4. Which period of an inferior planet is not observed from Earth?
   a. synodic
   b. sidereal
   c. [both periods are observed from Earth]
5. At which position would an inferior planet appear to be closest to Earth?
   a. superior conjunction
   b. inferior conjunction
   c. greatest elongation
   d. opposition
6. In what phase would an inferior planet be when it is closest to Earth?
   a. new
   b. full
   c. crescent
   d. gibbous
7. At which position would an inferior planet appear smallest?
   a. superior conjunction
   b. inferior conjunction
   c. greatest elongation
   d. opposition

8. At which position would an inferior planet appear dimmest?
   a. superior conjunction
   b. inferior conjunction
   c. greatest elongation
   d. [an inferior planet's brightness does not change]
9. When an inferior planet is full, it is also at
   a. its brightest.
   b. its dimmest.
   c. its closest.
   d. its farthest.
10. In which position is an inferior planet never seen?
   a. superior conjunction
   b. inferior conjunction
   c. greatest elongation
   d. opposition

OPEN-ENDED ACTIVITY

Using the skills you learned in this activity, try to determine a date when you will next be able to observe Venus from your location. Make a drawing of where Venus will be seen relative to the horizon. Record the date and time of the observation and the direction your drawing is facing.

VOYAGER: SKYGAZER CD-ROM ACTIVITIES

14
The Superior Planets

INTRODUCTION

The planets farther from the Sun than Earth, including Mars, are often called the *superior planets*. Their motion will be the focus of this activity.

PART 1: WHICH PLANETS ARE THE SUPERIOR PLANETS?

1. Select **The Solar System** under the **Explore** menu.
2. Which planets are superior? Record them on the RESULTS sheet.
3. Click on the **Time** arrow to watch the planets orbit the Sun. Record your observations on the RESULTS sheet.

PART 2: THE PERIODS OF MARS

File | Open Settings | Basics | Planet Orrery

1. Adjust the **Sun Distance** on the **Location Panel** to about 5 AU, or use the **Zoom** button until Mars is the outermost orbit in your Sky Chart. Use the **Planet Panel** under the **Control** menu to turn Earth and its **Name** on.
2. Set the time to today's date. Although the planets may be in different positions, your Sky Chart should look similar to Figure 14-1.
3. Using a 2-day time step, advance the time until Earth is between Mars and the Sun. When Earth is between the Sun and a superior planet, the configuration is called *opposition*. From Earth, the planet will appear opposite the Sun in the sky. Record the date of opposition on the RESULTS sheet.
4. Now advance the time until the Sun is between Earth and Mars. This configuration is called *superior conjunction*. Record the information requested on the RESULTS sheet.
5. Next, advance the time until Mars is again at opposition. Record the date of this opposition and determine the approximate *synodic period* of Mars. This is the time between two identical configurations between the planet, Earth, and the Sun.

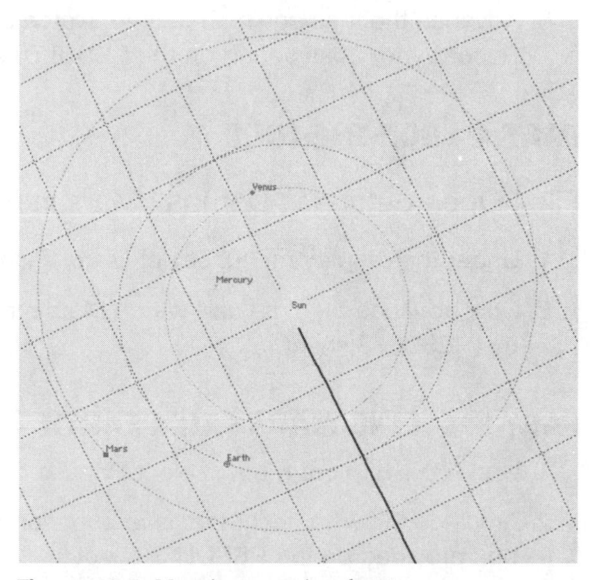

Figure 14-1 Mars is a superior planet

6. Take note of the position of Mars in its orbit around the Sun. Advance the time again, but this time ignore Earth. Stop when Mars has returned to approximately the same position in its orbit around the Sun that it was when you started. One of the gridlines in your Sky Chart may be a good starting and stopping point. The amount of time required for a planet to orbit the Sun is called its *sidereal period*. Determine and record Mars' approximate sidereal period.

PART 3: COMPARING THE PERIODS

1. You can observe the difference between a superior planet's synodic and sidereal period by selecting **Paths of the Planets** under the **Explore** menu. Select **Looping Path of Mars.**
2. Advance the time until Mars is at superior conjunction. In the **View from Earth** window, Mars will be lined up with the Sun (see Figure 14-2).

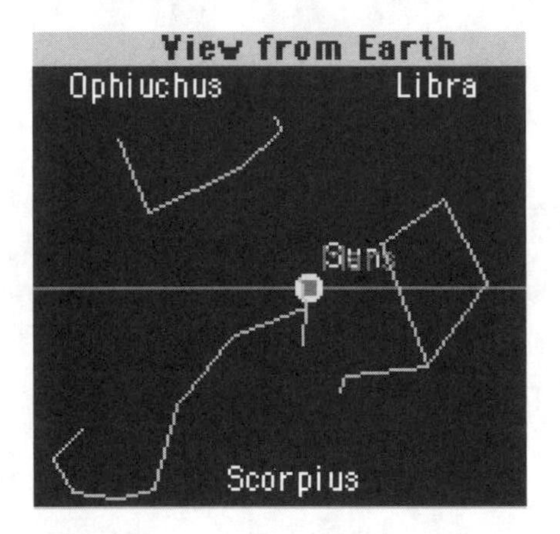

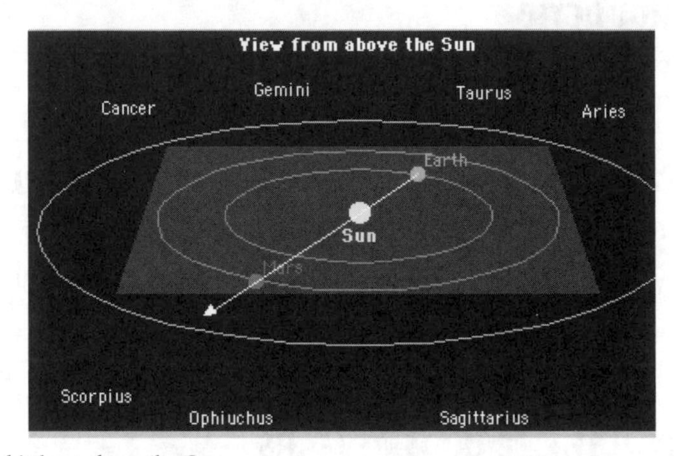

Figure 14-2 A view of Mars from Earth and its orbit from above the Sun

3. Now advance the time until Mars returns to approximately the same position in its orbit around the Sun (you can use whatever constellation Mars is "in" as a guide).
4. Is Mars at superior conjunction again? If not, continue to advance the time until you reach the next superior conjunction, and answer the questions on the RESULTS sheet.
5. Advance the time again. This time, watch carefully for what happens when Mars is at opposition. Record your observation on the RESULTS sheet.

PART 4: LOOP-THE-LOOP

File | Open Settings | Demos | Mars in Retrograde

1. Under the **Display** menu, select **Major Planets**, then **Scale Planets by 1×** to display Mars' symbol.
2. Now advance the time and watch Mars move through the stars. Record the required information on the RESULTS sheet.

PART 5: SEEING MARS FROM EARTH

1. Under the **Explore** menu, select **Phases of the Planets** and then **Phases of Mars**.
2. Advance the time until Mars is at superior conjunction. Record the phase of Mars and draw it in the box provided on the RESULTS sheet.
3. Now advance the time and observe the phases until Mars is at opposition. Record the phase at opposition on the RESULTS sheet.

PART 6: KEPLER'S THIRD LAW

File | Open Settings | Basics | Planet Orrery

1. Advance the time in 6-month steps to estimate the orbital (sidereal) periods of both Jupiter and Saturn. Use one of the gridlines in your Sky Chart as a starting and stopping point.
2. Three times (approximately equally spaced) during the orbit, stop and click on the planet to bring up its **Data Panel** and record the planet's **Sun Distance**. Record the required information in the table on the RESULTS sheet.

RESULTS SHEET 14 The Superior Planets

NAME ______________________ **DATE** ______________ **SECTION** ______________________

PART 1: WHICH PLANETS ARE THE SUPERIOR PLANETS?

Which planets are considered the superior planets?

The superior planets orbit the Sun CLOCKWISE | COUNTERCLOCKWISE *(circle one)*.

Do all the superior planets orbit in the ecliptic plane?

PART 2: THE PERIODS OF MARS

Start Date: ______________

Date of Mars' opposition: ______________

Date of Mars' superior conjunction: ______________

Date of Mars' next opposition: ______________

Synodic period of Mars = ______________ days

Synodic period from **Data Panel** = ______________ days

Sidereal period of Mars = ______________ days

Sidereal period from **Data Panel** = ______________ years × 365 = ______________ days

In Figure 14-3, draw Mars in its orbit at the positions of both opposition and superior conjunction.

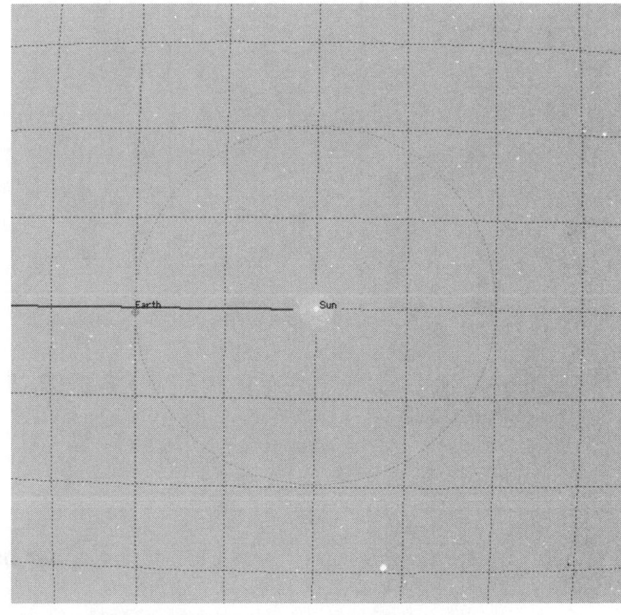

Figure 14-3 Draw Mars' position at both opposition and superior conjunction

PART 3: COMPARING THE PERIODS

In one sidereal period, does Mars return to the same point in its orbit relative to the Sun? YES | NO *(circle one)*

The SYNODIC | SIDEREAL *(circle one)* period of Mars is longer.

Does this agree with your results from Part 2? YES | NO *(circle one)*

When you viewed Mars from the Earth, what did you observe it do at opposition?

PART 4: LOOP-THE-LOOP

Draw the path of Mars you observed through the stars of Aquarius.

Figure 14-4 Draw Mars' path while in retrograde.

Date Mars began retrograde: _______________

Date retrograde ended: _______________

Duration of retrograde: _______________

PART 5: SEEING MARS FROM EARTH

In each box below draw the phase of Mars you observed.

Superior Conjunction	Opposition
Phase: _______________	Phase: _______________

What difference do you notice in the appearance of Mars in the two configurations? What is the reason for this difference?

In which of the configurations would Mars appear brighter? Explain your reasoning.

What differences do you notice between the phases of Mars and those of Venus?
(To review Venus's phases, go to **Explore | Phases of the Planets | Phases of Venus.**)

PART 6: KEPLER'S THIRD LAW

	Period (years)	S.D.* #1 (AU)	S.D. #2 (AU)	S.D. #3 (AU)	Average S.D. (AU)	P^2	a^3
Jupiter							
Saturn							

*S.D. = Sun Distance

Does Kepler's third law seem to hold for Jupiter and Saturn? What reason can you think of for any discrepancies?

Which of the superior planets seems to have the most eccentric orbit?

CONCLUSION

In the space below, write a conclusion for this activity. Briefly explain what you did and what you learned from it.

CHECK YOUR UNDERSTANDING 14: THE SUPERIOR PLANETS

MULTIPLE-CHOICE QUESTIONS

1. Which planet is not considered superior?
   a. Saturn
   b. Jupiter
   c. Mars
   d. [all of the above ARE superior]
2. In which phase will a superior NOT be observed?
   a. new
   b. full
   c. gibbous
   d. [actually, superior planets go through entire cycle phases]
3. Which period of a superior planet is longer?
   a. synodic
   b. sidereal
   c. [both periods are the same]
4. Which period of a superior planet is not observed from Earth?
   a. synodic
   b. sidereal
   c. [both periods are observed from Earth]
5. At which position is a superior planet in retrograde motion?
   a. superior conjunction
   b. inferior conjunction
   c. quadrature
   d. opposition
6. In which position will a superior planet appear full?
   a. superior conjunction
   b. opposition
   c. quadrature
   d. [two of the above]
7. At which position is a superior planet closest to Earth?
   a. superior conjunction
   b. inferior conjunction
   c. quadrature
   d. opposition

8. At which position should a superior planet appear brightest?
   a. superior conjunction
   b. inferior conjunction
   c. quadrature
   d. opposition
9. At which position should a superior planet appear dimmest?
   a. superior conjunction
   b. inferior conjunction
   c. quadrature
   d. opposition
10. In which position is a superior planet never seen?
    a. superior conjunction
    b. inferior conjunction
    c. quadrature
    d. opposition

OPEN-ENDED ACTIVITY

Using the skills you learned in this activity, try to determine a date of Mars' next opposition. Make a drawing of where Mars will be seen from your location. Record the date and time of the observation and the direction your drawing is facing. Include the horizon in your drawing.

Also attempt to record the beginning and end dates of Mars' retrograde motion during the opposition.

15
Observing the Planets

INTRODUCTION

The times that inferior and superior planets are visible from the Earth depend on what configuration they are in. In this activity you will focus on viewing Venus and Mars as an example of each type of planet, but what you will discover will work for all visible planets of each type.

PART 1: SEEING VENUS

File | Open Settings | Basics | Planet Orrery

1. Adjust the **Sun Distance** on the **Location Panel** to about 5 AU, or use the **Zoom** button until Mars is the outermost orbit. Bring up the **Planet Panel** under the **Control** menu, select Earth and its **Name** button, and deselect all the other planets and their **Names**, **Symbols**, and **Orbits** except Venus and the Sun. Your Sky Chart should look similar to Figure 15–1.
2. Now examine your Sky Chart. Also consider that Earth rotates counterclockwise as seen from above and that it is noon at a given location on Earth when it is *facing toward the Sun*. Try to predict when and where you would be able to see Venus when it is in the position in its orbit shown in your Sky Chart. Enter your prediction on the RESULTS sheet.
3. Now test your prediction. Change the view on the **Location Panel** from **Solar System** to **Earth**.
4. Deselect ALL the orbits on the **Planet Panel** and then select **Center Planet/Venus** under the **View** menu (or you can double-click on the Venus button on the **Planet Panel**).

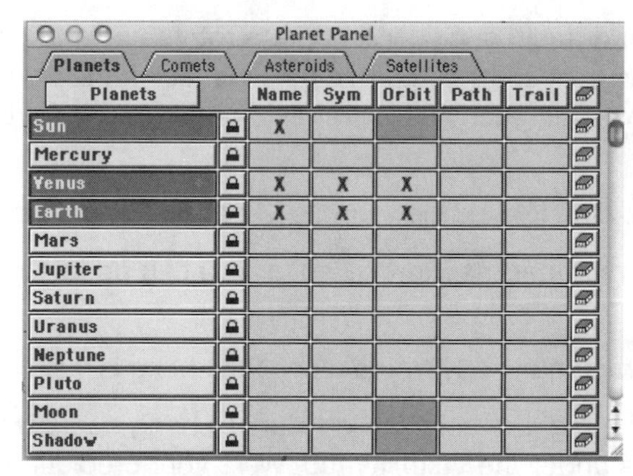

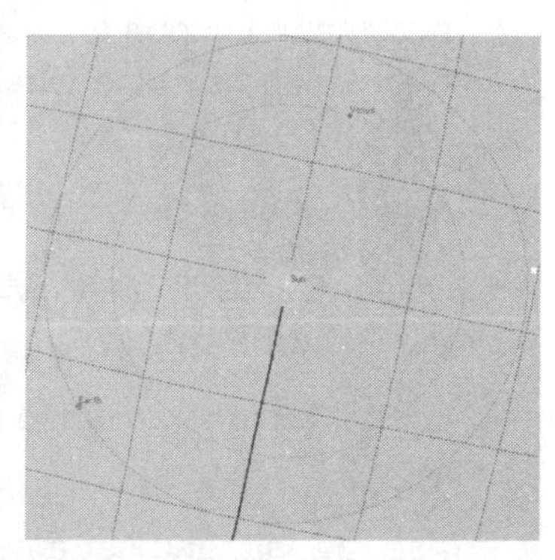

Figure 15-1 The **Planet Panel** and Sky Chart set for determining when is best for viewing Venus

5. Lock Venus by clicking on the **lock** button, and advance the time in 10-minute steps until Venus is visible. Remember that you can see a planet only if it is above the horizon at night.
6. Zoom out to 120° to get the big picture. When is Venus visible? Where is it in the sky? Were your predictions correct? Record this information on your RESULTS sheet.

PART 2: VENUS AND THE SUN

1. Now advance the time in 1-day steps and observe the change in Venus's position relative to the Sun. Stop when Venus is as close to the Sun as you can get it.
2. Alternately click on Venus and on the Sun to bring up their **Data Panels**. Record their **Distance** (from Earth—*not* **Sun Distance**!) under the **General** tab on the RESULTS sheet.
3. Also record the date of your observation.
4. Based on Venus being directly lined up with the Sun and on which object is farther away, make a drawing on the RESULTS sheet of the configuration of Earth, Venus, and the Sun.
5. You can check your prediction by again opening the setting below, but before doing this, *make sure you have recorded the date from the configuration on the* RESULTS *sheet*. Open the setting **File|Open Settings|Basics|Planet Orrery** and then set the time to the date you recorded on the RESULTS sheet. Follow the procedure in step 1 of Part 1 to check your prediction.

PART 3: THE EVENING STAR

1. Advance the time in 1-day steps until you think Venus will again be visible.
2. Record the position of Venus on Figure 15-4 on the RESULTS sheet and predict when and where you expect Venus to be visible.
3. To test your prediction, go back and repeat steps 4–7 of Part 1.

PART 4: MARS AT OPPOSITION

File | Open Settings | Basics | Planet Orrery

1. Adjust the **Sun Distance** on the **Location Panel** to about 5 AU, or use the **Zoom+** button until Mars is the outermost orbit. Bring up the **Planet Panel**, select Earth and its **Name** button, and deselect all the other planets and their **Names**, **Symbols**, and **Orbits** except Mars and the Sun.
2. Advance the time in 1-day steps until Mars is at *opposition*, opposite the Sun from Earth.
3. Examine your Sky Chart and again consider that Earth rotates counterclockwise as seen from above and that it is noon at a given location on Earth when it is facing the Sun. Try to predict when Mars will rise and set and when it will reach its highest altitude for the night. Enter your predictions on the RESULTS sheet.
4. Now test your predictions. Change the view on the **Location Panel** from **Solar System** to **Earth**.
5. Select **Center Planet | Mars** under the **View** menu (or you can double-click on the **Mars** button on the **Planet Panel**).
6. Lock Mars and advance the time in 10-minute steps until Mars is visible (above the horizon at night).
7. Zoom out to 120° to get the big picture. Click on Mars to bring up its **Data Panel** and check Mars' **rise**, **transit**, and **set** times under the **Visibility** tab. Were your predictions correct?

PART 5: MARS AT QUADRATURE

File | Open Settings | Basics | Planet Orrery

1. Follow the procedure in step 1 of Part 4. Then advance the time in 1-day steps until you get a configuration that looks similar to the one in Figure 15-2. This configuration is called *quadrature*. The lines from Earth to Mars and Earth to the Sun are 90° from each other.

2. Try to predict when Mars will be visible. Enter your prediction on the RESULTS sheet.
3. To test your prediction, follow the procedure in steps 4–7 of Part 4.
4. In Figure 15-5 on the RESULTS sheet, draw Mars at another position of quadrature and predict when it should be visible. Use *Voyager: SkyGazer* to help, if you want.

Figure 15-2 Determine when Mars is visible from quadrature

PART 6: OCCULTATION

File | Open Settings | Settings | Herschel Discovers Uranus

1. Use the **Time Panel** to move the time backward in 1-day steps until the Sun appears in front of, or *occults*, Uranus. Record this date and other requested information on the table on the RESULTS sheet. At which orbital configuration can an *occultation* of a superior planet by the Sun occur?
2. Advance the time until the Sun is again close to Uranus. Record this date and estimate the synodic period of Uranus. Bring up Uranus's **Data Panel** and check under the **Physical** tab to verify your result.
3. Now advance the time, watching Uranus's **Distance** (*not* **Sun Distance**) under the **General** tab on its **Data Panel**. Find and record the date that Uranus was closest to Earth. At which orbital configuration would this occur?
4. Uranus was discovered on March 13, 1781. Move time to this date and record the information requested on the table on the RESULTS sheet.

PART 7: TRANSIT

File | Open Settings | Settings | Venus Transit of 1769

1. When an inferior planet crosses the disk of the Sun, it is called a *transit*. Use the **Time Panel** to determine the duration of the 1769 transit of Venus.
2. Venus was also in transit on June 8, 2004. Use **Set Time** under the **Control** menu to change the date, the **Planet Panel** to center on and lock the Sun, and the **Time Panel** to change time in 1-minute steps to watch the 2004 transit. Record the duration of the 2004 transit of Venus as seen from the same location as in 1769. Record the requested information on the RESULTS sheet.

PART 8: CONJUNCTION

File | Open Settings | Demos | Planet Conjunction 1186

1. When two or more planets are close to one another in the sky, it is called a *conjunction*. Conjunctions of 2 or 3 planets are not uncommon. Gatherings of more than 3 planets are often called *massings.* How many planets were involved in the conjunction of 1186?
2. Search for the next conjunction that will be visible from your location. Set the location to your location and the time to the current date.
3. Select **Skyview | Star Atlas** under the **Chart** menu and click on **Zoom** and select **S360**. Also under the **Chart** menu deselect **Chart Labels | Star Names**.
4. Under the **Display** menu select **Reference Markers | Ecliptic Equator**. Use the **Planet Panel** to click off the images of Uranus, Neptune, Pluto, and the Moon.

5. Now advance time manually until you observe two or more planets appearing to overlap one another.
6. Change your **Skyview** under the **Chart** menu to **Local Horizon Coordinates**, and adjust your time to the nearest sunrise or sunset (you may have to try both) and click on the **Zoom** button to zoom into 30° to observe your conjunction. Record the information requested on the RESULTS sheet.

RESULTS SHEET 15 Observing the Planets

NAME ______________________ **DATE** ______________ **SECTION** ______________________

PART 1: SEEING VENUS

In the configuration you have set, when should Venus be visible from Earth? **AFTER SUNSET | MIDNIGHT | BEFORE SUNRISE** *(circle one)*.

In which part of the sky would you expect to see Venus? In the **WEST | SOUTH | EAST** *(circle one)*. Was your prediction correct? **YES | NO** *(circle one)*

PART 2: VENUS AND THE SUN

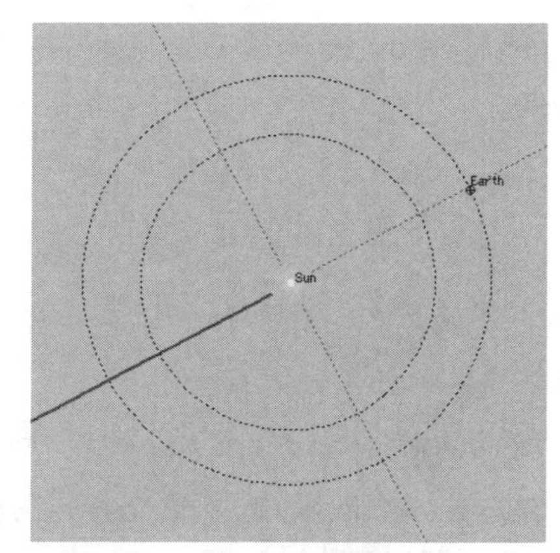

Figure 15-3 Draw the position of Venus in its orbit

Date: ______________

Distance to Sun: ______________

Distance to Venus: ______________

Based on which object is closer and the fact that they are nearly lined up from your point of view on Earth, draw the configuration of Earth, Venus, and the Sun on Figure 15-3.

Name of this configuration: ______________

Would you expect to be able to see Venus in this configuration? Why or why not?

Was your prediction correct? **YES | NO** *(circle one)* If not, record the position of Venus on Figure 15-3.

PART 3: THE EVENING STAR

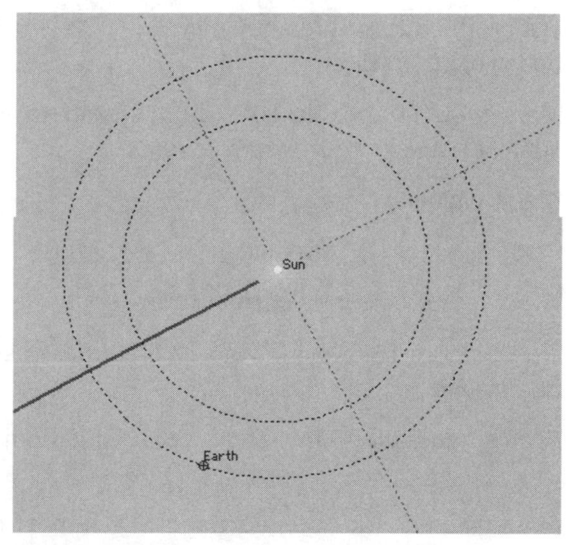

Figure 15-4 Draw Venus's position when it is called "The Evening Star"

In Figure 15-4, record the position Venus will be in, in its orbit around the Sun, when it becomes visible again.

In the configuration you have predicted, when should Venus be visible from Earth?

AFTER SUNSET | MIDNIGHT | BEFORE SUNRISE *(circle one)*

In which part of the sky would you expect to see Venus? In the **WEST | SOUTH | EAST** *(circle one)*

Was your prediction correct? **YES | NO** *(circle one)*

Would you ever expect to see an inferior planet at midnight? Why or why not?

PART 4: MARS AT OPPOSITION

During its opposition, when should Mars be visible from Earth?

Mars should rise around **SUNSET | MIDNIGHT | SUNRISE** *(circle one).*

Mars should reach its highest altitude (transit altitude) at about **SUNSET | MIDNIGHT | SUNRISE** *(circle one).*

Mars should set around **SUNSET | MIDNIGHT | SUNRISE** *(circle one).*

Mars' **Visibility** Data:

Rise: ___________ Transit: ___________ Set: ___________

Were your predictions correct? **YES | NO** *(circle one)*

Give two reasons that opposition would be the best time to view a superior planet.

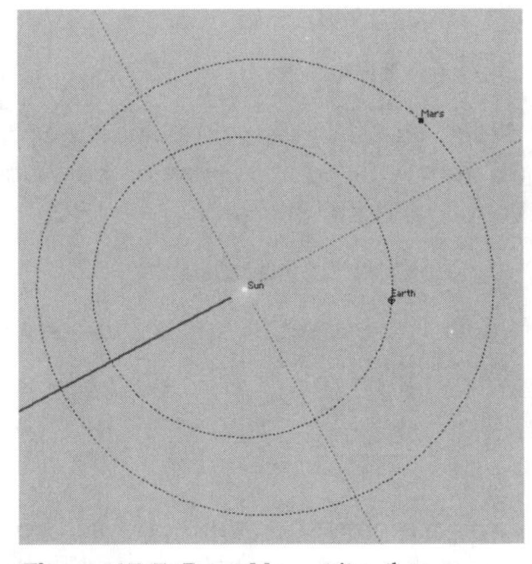

Figure 15-5 Draw Mars at its other quadrature

PART 5: MARS AT QUADRATURE

During the quadrature shown in Figure 15-5, when should Mars be visible from Earth?

Mars should be visible from about **SUNSET TO SUNRISE | MIDNIGHT TO SUNRISE | SUNSET TO MIDNIGHT** *(circle one).*

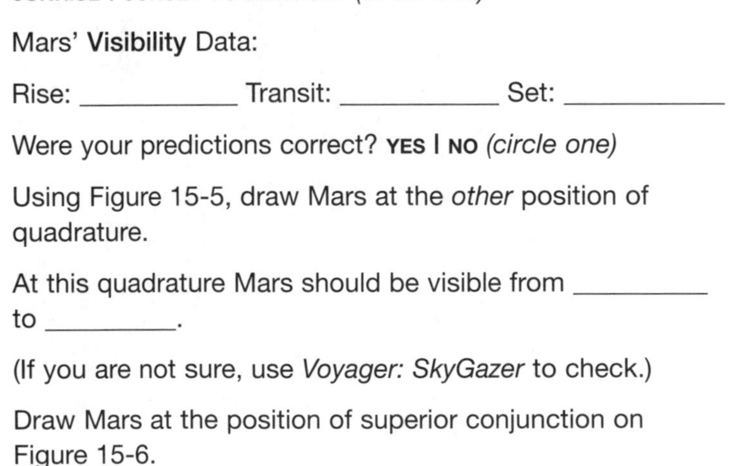

Mars' **Visibility** Data:

Rise: ___________ Transit: ___________ Set: ___________

Were your predictions correct? **YES | NO** *(circle one)*

Using Figure 15-5, draw Mars at the *other* position of quadrature.

At this quadrature Mars should be visible from _________ to _________.

(If you are not sure, use *Voyager: SkyGazer* to check.)

Draw Mars at the position of superior conjunction on Figure 15-6.

Would you expect to see a superior planet at superior conjunction in Earth's sky? Explain why or why not. (If you are not sure, use *Voyager: SkyGazer* to check.)

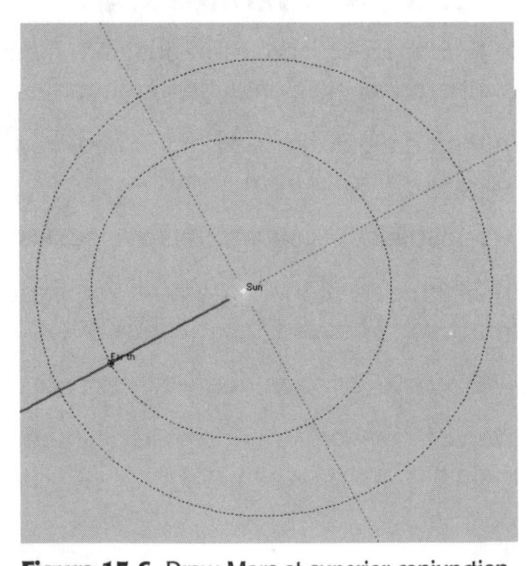

Figure 15-6 Draw Mars at superior conjunction

PART 6: OCCULTATION

Record the requested information for Uranus on the table below.

	Configuration	Date	Distance	Magnitude
1.				
2.				
3.				
4.	(Discovery)			

Synodic Period of Uranus: Your calculation: ___________ days

From Data Panel: ___________ days

On which date was Uranus brightest?

Was Uranus at its brightest when discovered?

Does Uranus's brightness change much through its synodic period? Explain your answer.

PART 7: TRANSIT

Draw the path of Venus across the disk of the Sun for both transits and record the duration (the amount of time Venus was in front of the Sun's disk) of each transit.

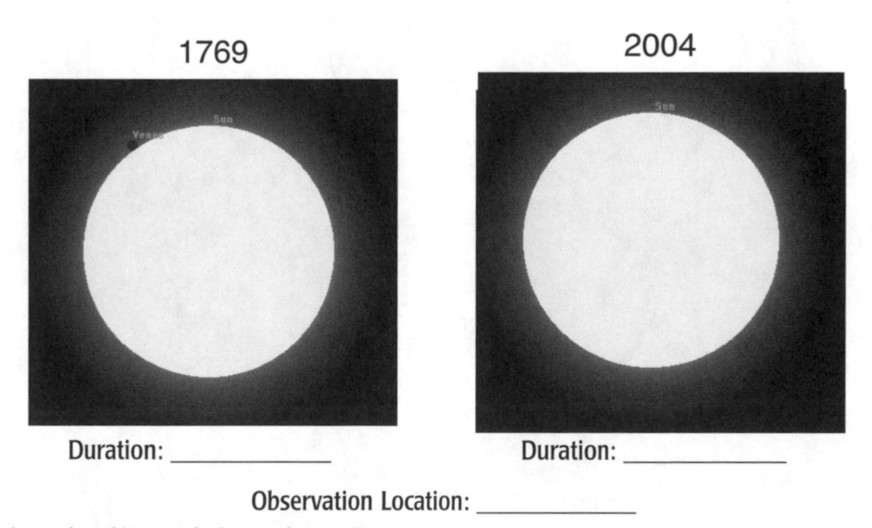

Duration: ___________ Duration: ___________

Observation Location: ___________

Figure 15-7 Draw the paths of Venus during each transit

A transit by an inferior planet occurs at SUPERIOR | INFERIOR *(circle one)* conjunction.

PART 8: CONJUNCTION

Planets involved in 1186 conjunction: ___________

Planets involved in your conjunction: ___________

Date and time observed: ___________

CONCLUSION

In the space below, write a conclusion for this activity. Briefly explain what you did and what you learned from it.

CHECK YOUR UNDERSTANDING 15: OBSERVING THE PLANETS

MULTIPLE-CHOICE QUESTIONS

1. In which position will Venus appear to be farthest away from the Sun in the sky?
   a. superior conjunction
   b. inferior conjunction
   c. greatest elongation
   d. [Venus always appears the same distance from the Sun]
2. When at a greatest elongation, when will Venus be best seen?
   a. at midnight
   b. early evening after sunset
   c. early morning before sunrise
   d. [either a or b, depending on whether the elongation is western or eastern]
3. In which position will Venus appear to be closest to the Sun in our sky?
   a. superior conjunction
   b. inferior conjunction
   c. greatest elongation
   d. [it will appear very close to the Sun in either a or b]
4. At which time is Venus never seen?
   a. early evening after sunset
   b. early morning before sunrise
   c. midnight
   d. [Venus could be seen at any of these times]
5. At which position could Venus be called a “Morning” or “Evening” star?
   a. superior conjunction
   b. inferior conjunction
   c. greatest elongation
   d. [it will appear very close to the Sun in either a or b]
6. At which time can Mars never be seen?
   a. early evening after sunset
   b. early morning before sunrise
   c. midnight
   d. [Mars could be seen at any of these times]

7. From which position would Mars be visible all night?
   a. superior conjunction
   b. inferior conjunction
   c. quadrature
   d. opposition
8. When at quadrature, Mars would be seen
   a. all day.
   b. all night.
   c. both day and night.
   d. [Mars would not be seen at all].
9. When the disk of a planet passes in front of a star, it is called a (an)
   a. eclipse.
   b. transit.
   c. occultation.
   d. [none of the above].
10. In which position would Venus be during a transit in front of the Sun?
    a. superior conjunction
    b. inferior conjunction
    c. greatest elongation
    d. [either a or b]

OPEN-ENDED ACTIVITY

Using the skills you learned in this activity, try to determine which planets are currently visible from your location at night. Record the times between which they will be visible.

16

Asteroids, Comets, and Meteors

INTRODUCTION

Asteroids, comets, and meteors are often collectively referred to as the debris of the solar system. Despite that distinction, there are many interesting observations that can be made involving these types of objects, some of which you will simulate in this activity.

PART 1: ASTEROIDS

File | Open Settings | Basics | Planet Orrery

1. Zoom in or set the **Sun Distance** on the **Location Panel** to about 15 AU so that Jupiter is the outermost orbit in your Sky Chart.
2. Bring up the **Planet Panel** under the **Control** menu and select the **Asteroids** tab. Then click the **Asteroids** button off, then on again to turn on the asteroids. Also make sure that their **Names**, **Symbols**, and **Paths** are on. Your **Planet Panel** should look similar to Figure 16-1. Under the **Comets** tab, use the **Comets** button to turn off any comets that you see.
3. Click on each asteroid to bring up its **Data Panel** and record the information on the table on the RESULTS sheet. Record the asteroids in order of their Asteroid number (this number is the order in which they were discovered).
4. Change the time step on the **Time Panel** to 7 days. Set the time to the current date, then advance the time and watch the asteroids orbit the Sun.
5. Use the **Ecliptic Latitude** on the **Location Panel** to view the orbits from above, 90° latitude, and the side, 0° latitude. Answer the questions on the RESULTS sheet.

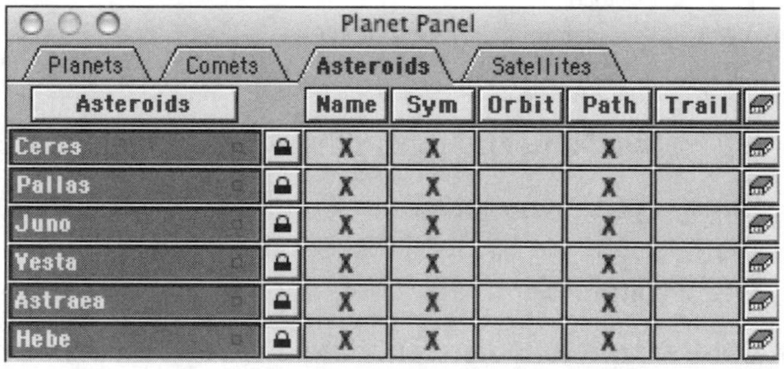

Planet Panel

Planets / Comets / Asteroids / Satellites

Asteroids		Name	Sym	Orbit	Path	Trail
Ceres		X	X		X	
Pallas		X	X		X	
Juno		X	X		X	
Vesta		X	X		X	
Astraea		X	X		X	
Hebe		X	X		X	

Figure 16-1 The **Planet Panel** with the Asteroids tab selected

PART 2: ASTEROIDS AND KEPLER'S THIRD LAW

Use the table on the RESULTS sheet to verify Kepler's third law.

PART 3: THE DISCOVERY OF CERES

1. Select **Earth** on the **Location Panel** and use the location marker ° to change your location to northern Italy, as shown in Figure 16-2. This is where the monk Guiseppe Piazzi first spotted Ceres' apparent motion in front of the background stars on December 31, 1800.
2. Use **Set Time** under the **Control** menu to change the time to midnight on the day of discovery, and set your **Sky View** under the **Chart** menu to **Star Atlas Coordinates.**
3. Then under the **View** menu select **Center on Asteroid/Ceres**, and zoom in to 30°.
4. Use the **Planet Panel** to make sure that Ceres' **Name**, **Symbol**, and **Path** are all on.
5. Advance the time in 1-day steps to observe Ceres' path through the stars for about the next month. Draw the path on the figure on the RESULTS sheet and answer the questions.

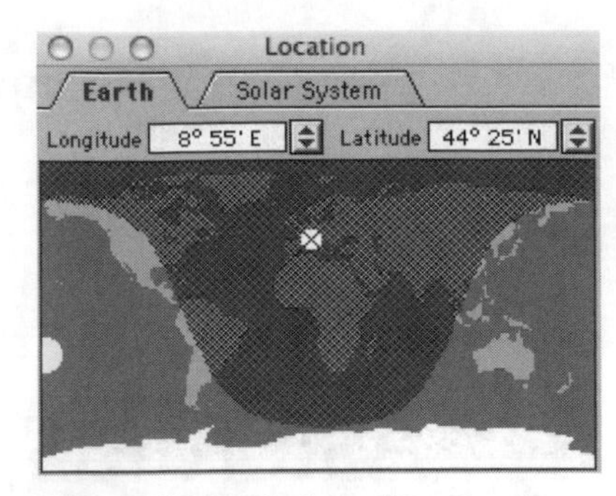

Figure 16-2 The **Location Panel** with the location marker set in northern Italy

PART 4: MORE ASTEROIDS

File | Open Settings | Basics | Planet Orrery

Alternately select each **Asteroid Group** under the **Display** menu, then complete the table and answer the questions on the RESULTS sheet. In the last column of the table, try to describe what the group has in common; for instance, the **Main Asteroid Belt** is comprised of asteroids located mostly between the orbits of two specific planets. You will have to use the **Zoom** button to move in and out to find each **Asteroid Group**.

PART 5: COMETS

File | Open Settings | Basics | Planet Orrery

1. Select the **Comet** button on the **Planet Panel**. Turn on the **Name**, **Symbol**, and **Path** of each comet.
2. Bring up each comet's **Data Panel.** From the **General** tab, record its name and **Sun Distance** on the table on the RESULTS sheet. Using the **Physical** tab, record each comet's **Perihelion Date** and **Orbital Period.** Then find photographs of the comets under the **Pictures** tab.
3. Set the time to the current date, then advance the time and watch the comets orbit the Sun.
4. Change the **Ecliptic Latitude** on the **Location Panel** to view the orbits from above (90° latitude) and the side (0° latitude). Answer the questions on the RESULTS sheet.

PART 6: COMET HYAKUTAKE

File | Open Settings | Demos | Hyakutake at Perihelion

1. Advance the time and watch Comet Hyakutake's orbit as it enters and leaves the inner solar system.
2. Stop the time when Hyakutake leaves your Sky Chart. Adjust the **Ecliptic Longitude** on the **Location Panel** to view the comet's path from various perspectives.
3. Set the latitude to 0° and the longitude to 180° to view the orbit edge–on. Zoom in and draw this view of the orbit in the figure on the RESULTS sheet, then estimate the inclination angle (the angle that the comet's orbit makes with the *plane of the ecliptic*, the plane defined by the Earth's orbit around the Sun).

4. When Hyakutake was near its perihelion, was it above or below the plane of the ecliptic? What part of the sky would you expect it to be in when observed from Earth?

File | Open Settings | Demos | Hyakutake nears Earth

5. Under the **Control** menu, Set **Time** to March 15, 1996, at 12:00 A.M.
6. Under the **Chart** menu, select **Sky View|Local Horizon Coordinates** and under the **View** menu, select **Center Comet|Hyakutake**.
7. Advance the time in 1-day steps and observe the changing position of the comet until about March 31. The majority of the United States experienced a view similar to this.
8. Is the comet in the part of the sky you expected based on your answer to the last question? To what reference point in the sky did it come very close? On the RESULTS sheet draw the comet on the night its tail had its longest length on the figure and answer the questions.

PART 7: COMET HALE-BOPP

File | Open Settings | Basics | Orbit of Hale-Bopp

1. Zoom in to 10° and advance the time to observe the orbit of Comet Hale-Bopp. Pay special attention to the changing length of the comet's tail and the direction it is pointing. Answer the questions on the RESULTS sheet.

File | Open Settings | Settings | Hale-Bopp at 8PM

2. Advance the time in 1-day steps to April 1, 1997, and change the local time to 9:00 P.M. Zoom in to 60° and draw Comet Hale-Bopp on the figure on the RESULTS sheet.

PART 8: METEORS

1. Use the **location marker** ⊕ on the **Location Panel** to go to your approximate location. Make sure your **Sky View** under the **Chart** menu is set to **Local Horizon Coordinates**.
2. Make sure that the meteor button on the **Display Panel** is on (highlighted). 
3. Under the **Display** menu, select **Meteor Showers** and a list of meteor showers, shown in Figure 16-3 will appear.
4. Highlight a meteor shower and click the **Show Shower** button to see the shower.
5. Answer the questions on the RESULTS sheet.

Meteor Showers

Shower	Date	Rate	Duration
Quadrantids	Jan 3	90 per/hour	6 days
Lyrids	Apr 22	15 per/hour	6 days
Eta Aquarids	May 6	30 per/hour	10 days
Capricornids	Jul 24	8 per/hour	30 days
Delta Aquarids	Jul 29	20 per/hour	30 days
Iota Aquarids	Aug 6	10 per/hour	20 days
Perseids	Aug 12	90 per/hour	25 days
Orionids	Oct 22	25 per/hour	20 days
Taurids	Nov 5	15 per/hour	30 days
Leonids	Nov 17	12 per/hour	10 days
Geminids	Dec 14	90 per/hour	8 days
Ursids	Dec 23	8 per/hour	8 days

Cancel   Show Shower

Figure 16-3 Meteor showers

RESULTS SHEET 16 Asteroids, Comets, and Meteors

NAME ______________________ **DATE** ____________ **SECTION** ______________

PART 1: ASTEROIDS

In the table below, fill out the information from the various tabs on the asteroids' **Data Panels**.

Asteroid #	Name	Diameter	Sun Distance	Semimajor Axis	Eccentricity	Orbital Period
1						
2						
3						
4						
5						
6						

Which asteroid is the largest?

Which asteroid is farthest from the Sun?

Closest to the Sun?

Having watched their orbits, do you think the two asteroids that were your answers to the last two questions will always be correct? Explain your answer.

Which asteroid's orbit is most inclined to the ecliptic plane?

The orbits of all these asteroids are between the orbits of the planets ________ and ________.

PART 2: ASTEROIDS AND KEPLER'S THIRD LAW

Use the **Orbital Period** and **Semimajor Axis** of each asteroid in the table in Part 1 to verify Kepler's third law, $P^2 = a^3$.

Asteroid #	Name	Orbital Period (P)	Semimajor Axis (a)	P^2	a^3
1					
2					
3					
4					
5					
6					

PART 3: THE DISCOVERY OF CERES

Figure 16-4 shows the position of Ceres on December 31, 1800, the day it was discovered. Draw a line that represents Ceres' motion in front of the background stars over the month before and after its discovery.

How does the fact that Ceres' path is changing direction tell us that this was a likely time for it to be discovered?

Which planet passed through the field of view?

Figure 16-4 Draw Ceres' path in front of the background stars at the time of its discovery

PART 4: MORE ASTEROIDS

In the table below, fill out the name of each major **Asteroid Group** and what the members of the group have in common.

Asteroid Group	Common Factor

PART 5: COMETS

On the table below, fill out the information from the comets' **Data Panels**.

Comet	Date of Last Perihelion	Sun Distance	Orbital Period

Which comet had the most recent perihelion?

Which comet is farthest from the Sun?

Having watched their orbits, does it look like any of these comets will be returning to the inner solar system any time soon? Which comet will return first?

Does looking at the orbital periods support your last answer?

In Figure 16-5, draw the complete orbit of the comet that will be the first to return to the inner solar system. Also, mark on its orbit approximately where this comet is right now and answer the following questions.

How far is the comet from the Sun right now?

Between what two planets' orbits is the comet located?

How far is the comet's aphelion from the Sun?

When will the aphelion occur?

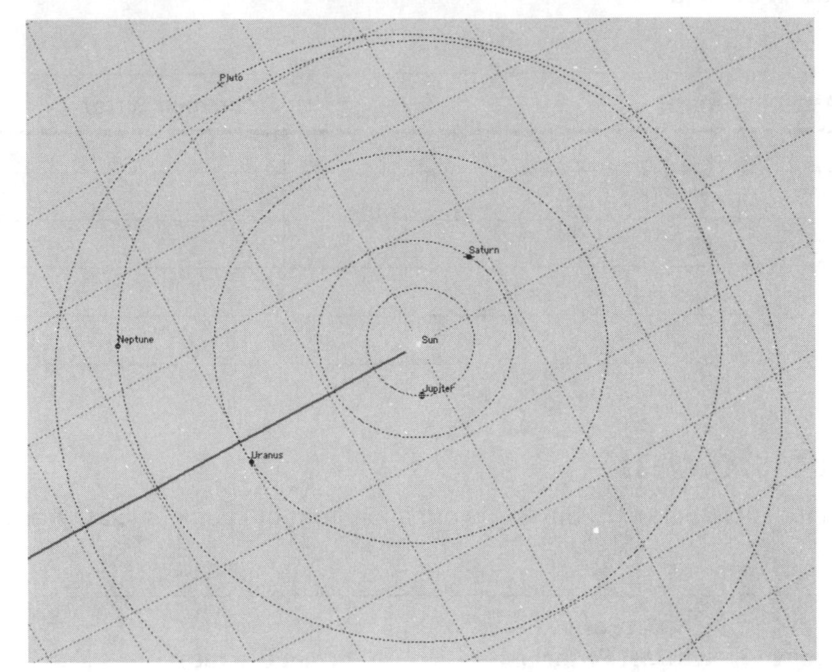

Figure 16-5 Draw the entire orbit of the comet that will return to the inner solar system first

As it gets closer to the Sun, what noticeable change do you see in the comet's motion?

PART 6: COMET HYAKUTAKE

In the space below, draw the plane of the orbit of Comet Hyakutake relative to the plane of the ecliptic. Draw the plane of the ecliptic as a horizontal line.

Estimated angle of inclination of orbit: ________ °

Hyakutake's perihelion was ABOVE | BELOW *(circle one)* the plane of the ecliptic.

When viewed from the Earth, in which part of the sky would you expect to have seen Comet Hyakutake?

In Figure 16-6, draw the comet in its position at the time its tail was longest. What was the date? __________

Were you correct about the part of the sky in which the comet would be seen?

What reference point in the sky was the comet very close to? Label this point in Figure 16-6.

Figure 16-6 Draw Comet Hyakutake as it looked on the date that its tail was longest

PART 7: COMET HALE-BOPP

After watching the orbit of Comet Hale-Bopp, explain what determines the length of a comet's tail and the direction in which it points.

In Figure 16-7, draw the position of Comet Hale-Bopp on April 1, 1997, at 9 P.M.

Figure 16-7 Draw Comet Hale-Bopp as it was seen on April 1, 1997

PART 8: METEORS

Which meteor shower(s) have the greatest rate?

Which meteor shower(s) have the longest duration?

Which shower has both a high rate and a long duration?

Click on **Show Shower** to observe this shower.

Can you figure out where the shower gets its name? *Hint:* Click on some of the stars near where the meteors are coming from and look at what constellation they are in.

Can you think of another reason, besides rate and duration, that this is the most popular of meteor showers, even though some others have similar rates and duration?

CONCLUSION

In the space below, write a conclusion for this activity. Briefly explain what you did and what you learned from it.

CHECK YOUR UNDERSTANDING 16: ASTEROIDS, COMETS, AND METEORS

MULTIPLE-CHOICE QUESTIONS

1. The main belt of asteroids is located between __________ and __________.
   a. Earth; the Sun
   b. Earth; Mars
   c. Mars; Jupiter
   d. Neptune; Pluto
2. Which DOES NOT appear to be a property of asteroids?
   a. much smaller than planets
   b. more eccentric orbits than planets
   c. orbits of higher inclination than planets
   d. [all seem to be property of asteroids]
3. How are asteroids discovered?
   a. they visually appear much brighter than nearby stars
   b. through apparent motion in front of the background stars
   c. they can be seen when they cross planetary orbits
   d. [all of the above are methods used to discover asteroids]
4. An asteroid with a period of eight years is most likely a (an)
   a. member of the main belt.
   b. Aten.
   c. Apollo.
   d. Centaur.
5. Which asteroid group's members seem to average the farthest distance from the Sun?
   a. the main belt
   b. Aten
   c. Apollo
   d. Centaur
6. Which objects usually have the most eccentric orbits?
   a. asteroids
   b. comets
   c. meteors
   d. planets

7. Which comet is most likely to return to the inner solar system first?
   a. LINEAR
   b. Hale-Bopp
   c. Hyakutake
   d. Ikea-Zhang
8. Which meteor shower has the longest duration?
   a. Eta Aquarids
   b. Persieds
   c. Orionids
   d. Geminids
9. Which meteor shower has the lowest rate?
   a. Quadrantids
   b. Persieds
   c. Orionids
   d. Geminids
10. Which is a likely reason for the popularity of the Persieds?
    a. summer date
    b. high rate
    c. long duration
    d. [probably all of the above]

OPEN-ENDED ACTIVITY

Using the skills you learned in this activity, determine the date of the next meteor shower and its rate and duration. Also determine whether the meteor shower will be visible from your location. Explain how you determined this.

17
Satellites and Spacecraft

INTRODUCTION

For nearly 50 years now, we have been reaching farther and farther out into space through the use of satellites and spacecraft. In this activity you will join some of the recent and current explorers of our solar system (and beyond) in their journeys.

PART 1: SATELLITES

File | Open Settings | Satellites March 24.02

1. Use the **Location Panel** to move to your location. Use the **Time Panel** to set the time to the current date at sunset.
2. Now advance the time in 1-minute steps to observe satellites that appear above your horizon. Count how many satellites become visible during the entire night. Use the **Data Panel** of several satellites that cross the sky high overhead (through the middle of your Sky Chart) to find their names. Record the information requested on the RESULTS sheet.

PART 2: THE HUBBLE SPACE TELESCOPE

File | Open Settings | Demos | Tracking Hubble – March 2002

1. Click on the Hubble Space Telescope (HST) to bring up its **Data Panel**. Record the information requested on the RESULTS sheet.
2. Change to your location (do not change the date), and advance time to see if the HST becomes visible (above your horizon). If it does, record when, for about how long, and approximately where to look for it.

PART 3: SPACECRAFT

File | Open Settings | Spacecraft | Galileo to Jupiter

The **Spacecraft** folder is in the *SkyGazer* Folder, with the **Basics, Demos**, and **Settings** folders you have been using.

1. Advance the time and use the setting to answer the questions on the RESULTS sheet. Click on the objects encountered by *Galileo* to bring up their **Data Panels** for help.

File | Open Settings | Spacecraft | Voyager I and Voyager II

2. Use the **Planet Panel** to turn the **Planet symbols** on, then advance the time and use the setting to answer the questions on the RESULTS sheet.

3. Use the other settings in the **Spacecraft** folder to answer the questions on the RESULTS sheet.

PART 3: SCIENCE FICTION

File | Open Settings | Settings | A Trip to Antares

1. Advance the time to run the setting, then answer the questions on the RESULTS sheet.

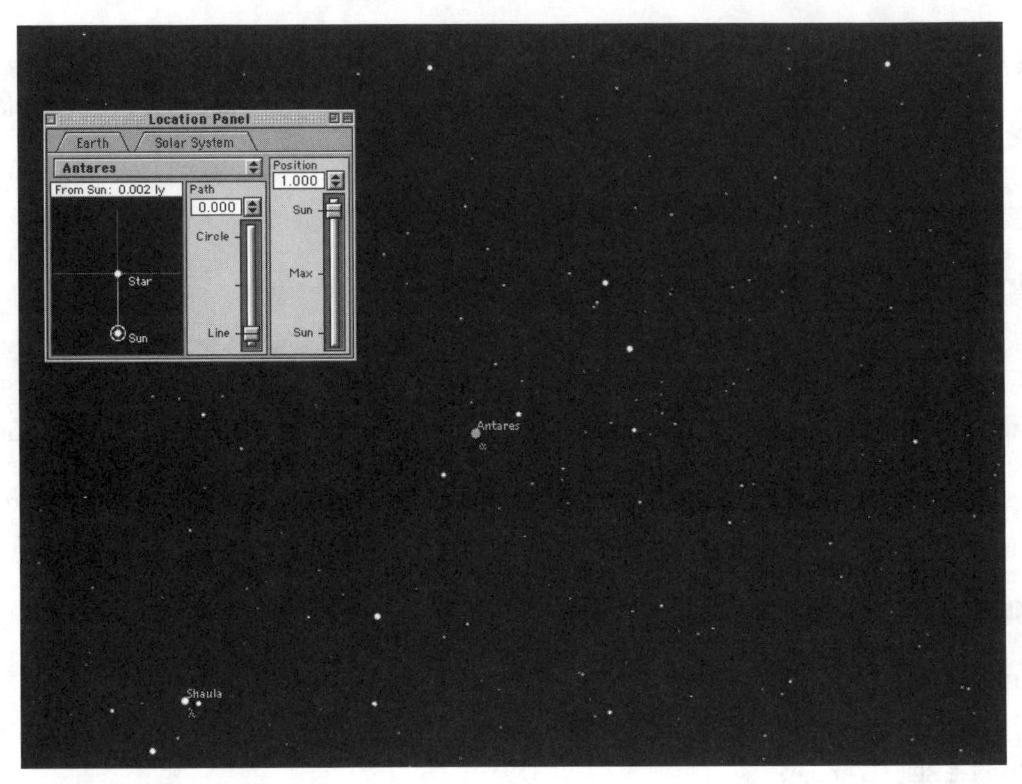

Figure 17-1 A trip to Antares

RESULTS SHEET 17 Satellites and Spacecraft

NAME ______________________ **DATE** ____________ **SECTION** ______________

PART 1: SATELLITES

Number of satellites observed in a night: ____________

Record the information about a satellite that you observed high overhead.

Your Location:____________ Date:____________

Satellite Name:____________

Rise Time:____________ Set Time:____________

Current Altitude: ____________ km (height above the Earth's surface)

Mean Motion: ____________ revs per day

Use the *Mean Motion* (in revolutions per day) to determine the satellite's orbital period in minutes:

$$\text{period} = \frac{1}{\textit{rev/day}} \times \frac{\textit{24 hours}}{\textit{day}} \times \frac{\textit{60 minutes}}{\textit{hour}} = \underline{\hspace{3cm}} \text{ minutes}$$

PART 2: THE HUBBLE TELESCOPE

Current Altitude: ____________ km

Mean Motion: ____________ revs per day

Period = ____________ minutes*

*calculate just like in Part 1

Was the HST visible from your location **YES | NO**

If it was visible:

Rise Time: ________

Set Time: ________

Time Visible: ________

In what part of the sky was it visible?

PART 3: SPACECRAFT

Galileo

What year was *Galileo* launched?

How many times did *Galileo* return to fly by Earth? In what years?

Which asteroids did *Galileo* fly by? In what years?

Label each picture in Figure 17-2 with the asteroid's name:

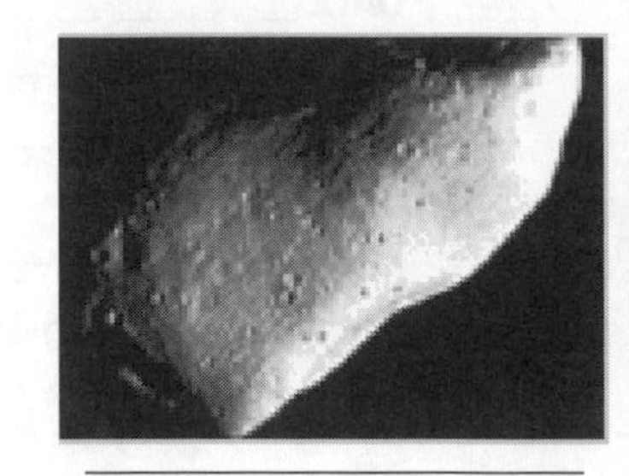

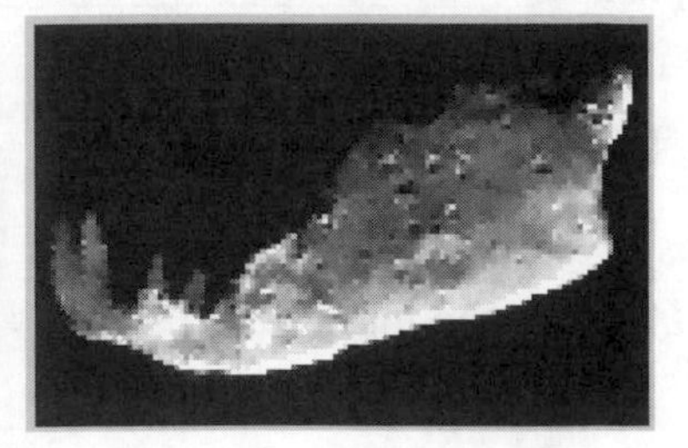

Figure 17-2 Label both asteroids with their correct names

What was *Galileo's** ultimate destination? When did she reach it?

Voyager I and Voyager II

In what year were the *Voyagers* launched?

In what year did they encounter Jupiter? Saturn?

Which *Voyager* went on beyond Saturn?

How long did it take to get from Saturn to Uranus? From Uranus to Neptune?

A radio message traveling at the speed of light would take 8 minutes to travel 1 AU the distance from the Earth to the Sun. When *Voyager* was at Neptune, approximately how long did it take to radio pictures back to the Earth?

Advance the time to the current date. Click on the end of the **Path** of each *Voyager* to bring up its **Data Panel** and record its Sun Distances. How far away are they from home? How long would it take to send a radio message now?

Note: Galileo's mission is now over.

Other Spacecraft

What did the *Giotto* spacecraft explore? When?

Besides *Voyager* and *Galileo*, what other spacecraft visited Jupiter?

What did the *Ulysses* spacecraft explore? When?

PART 3: SCIENCE FICTION

Click on Antares to bring up its **Data Panel** and record its **Distance.**

Distance to Antares: ______________ light-years.

Light itself travels $c = 3 \times 10^8$ m/s = 186,000 miles/sec = 1 light-year per year.

So, if your starship could really travel to Antares in one year, how fast would it have to be going? Do you think this is possible? Why or why not?

Speed of your starship = _____________ $\times c$

CONCLUSION

In the space below, write a conclusion for this activity. Briefly explain what you did and what you learned from it.

CHECK YOUR UNDERSTANDING 17: SATELLITES AND SPACECRAFT

MULTIPLE-CHOICE QUESTIONS

1. A satellite orbits the Earth once in about
   a. 1.5 minutes.
   b. 15 minutes.
   c. 1.5 hours.
   d. 15 hours.
2. Which spacecraft was launched first?
   a. *Galileo*
   b. *Giotto*
   c. *Ulysses*
   d. *Voyager I*
3. Which spacecraft was launched most recently?
   a. *Galileo*
   b. *Giotto*
   c. *Ulysses*
   d. *Voyager I*
4. Which spacecraft did NOT visit Jupiter?
   a. *Pioneer*
   b. *Voyager II*
   c. *Giotto*
   d. *Galileo*
5. Which spacecraft explored beyond Saturn?
   a. *Voyager I*
   b. *Voyager II*
   c. *Giotto*
   d. *Ulysses*
6. How did the *Voyager I*'s mission end?
   a. It disappeared at Jupiter.
   b. It spiraled up out of the solar system after visiting Saturn.
   c. It visited Pluto.
   d. It just disappeared.

7. How did the *Voyager II*'s mission end?
   a. It never even made it to Jupiter.
   b. It disappeared at Saturn.
   c. It visited Pluto.
   d. It left the solar system and is still going.
8. Which spacecraft visited Comet Halley?
   a. *Galileo*
   b. *Giotto*
   c. *Ulysses*
   d. *Voyager II*
9. Which spacecraft studied the Sun?
   a. *Galileo*
   b. *Giotto*
   c. *Ulysses*
   d. *Voyager II*
10. "A Trip to Antares" is science fiction because
    a. Antares is so far away.
    b. of how fast the starship traveled.
    c. [both of the above].
    d. [none of the above].

OPEN-ENDED ACTIVITY

Use the skills you learned in this activity to try and find if and when the International Space Station will be visible from your location. Record the time and date that you found it. *Hint:* Look under the **View** menu.

18
The Solar System Planets, Moons, and Rings

INTRODUCTION

In this activity you will use the physical and orbital characteristics of planets, moons, and rings in our solar system to make comparisons between them.

PART 1: PLANETARY DATA

File | Open Settings | Basics | Planet Orrery

1. Adjust the **Sun Distance** on the **Location Panel** to about 5 AU, or use the **Zoom+** until Mars is the outermost orbit. Use the **Planet Panel** to turn the Earth's symbol on. Alternately, click on each of these inner planets to bring up its **Data Panel**. Use the different tabs to record all the information requested on the table on the RESULTS sheet. Adjust the **Sun Distance** on the **Location Panel** back out to about 100 AU or zoom out, and record the information for the outer planets.
2. Once you have filled in all the data on the table, answer the questions on the RESULTS sheet.

PART 2: PLANET TYPES

1. Use the same setting as in Part 1. Alternately, click on each of the inner planets to bring up its **Data Panel**. Examine the pictures under the **Picture** tab for each of the terrestrial planets. Based on the pictures, write a short description on the RESULTS sheet of each planet's surface. Focus on the apparent differences between the surfaces.
2. Once you have written your descriptions, answer the questions on the RESULTS sheet.

PART 3: JOVIAN MOONS

File | Open Settings | Settings | Three Moons on Jupiter

1. Zoom out until you can see the orbits of all the moons shown on the Sky Chart.
2. Click on each moon you see to bring up its **Data Panel**. Use the various tabs to find the information requested on the table on the RESULTS sheet and record the information.
3. Use the settings below and repeat steps 1 and 2 above to record information for the moons of Saturn, Uranus, and Neptune on the RESULTS sheet.

File | Open Settings | Settings | Locked on Dione
File | Open Settings | Settings | Moons of Uranus
File | Open Settings | Settings | Neptune and Triton

PART 4: LARGE MOONS

File | Open Settings | Settings | Moon Around the Earth

1. Click on the Moon to bring up its **Data Panel** and record its **Diameter**, then calculate $Diameter_{Moon}$ $Diameter_{Earth}$. Record the information on the RESULTS sheet.

File | Open Settings | Settings | Pluto and Charon

2. Zoom in until Charon becomes visible. Click on Charon to bring up its **Data Panel** and record Charon's **Diameter**, then calculate $Diameter_{Charon}$ / $Diameter_{Pluto}$. Record the information on the RESULTS sheet.

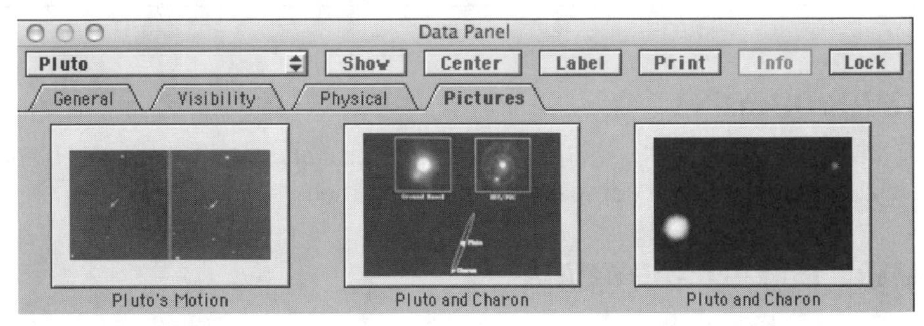

Figure 18-1 Pluto and Charon

PART 5: RINGS

File | Open Settings | Basics | Saturn and 28 Sgr

1. Advance the time manually in 1-minute steps to observe the occultation of the star 28 Sagittarius (28 Sgr) by Saturn. On the RESULTS sheet, record the time that 28 Sgr was first occulted by the rings, the time that it was first occulted by the planet, the time it emerged from behind the planet, and the last time it emerged from behind the rings.
2. Dividing the time that the star was behind Saturn only by the total time it was occulted will give the ratio of the diameter of the ring system to the diameter of the planet. This should be less than the planet's *Roche Limit*, which is a boundary between the ring system and the moons.

Figure 18-2 Saturn and its moon Tethys

3. The Roche Limit can be calculated from the formula given on the RESULTS sheet. Bring up Saturn's **Data Panel** and the **Data Panel** for the closest moon of Saturn that you can find in your Sky Chart. Record the density of Saturn and the moon from the **Physical** tab on the **Data Panel**. Calculate the Roche Limit with the formula and see how it compares with your result from step 2.

RESULTS SHEET 18 The Solar System Planets, Moons, and Rings

NAME ______________________ **DATE** ______________ **SECTION** ______________

PART 1: PLANETARY DATA

Planet	(km)	Mass Diameter	Density Earth = 1	Sun Water = 1 (AU)	Orbital Distance (years)	Rotation Period	Moons (days)
Mercury							
Venus							
Earth							
Mars							
Jupiter							
Saturn							
Uranus							
Neptune							
Pluto							

Based on your data in the table above:

Describe the properties that the Earth-like (or terrestrial) planets have in common. Which other planets fit into this category?

Describe the properties that the Jupiter-like (or jovian) planets have in common. Which other planets fit into this category?

Which planet does not seem to fit either category?

What do the densities of the terrestrial planets suggest about them? The jovian planets?

PART 2: PLANET TYPES

Describe the planetary surfaces of the following planets from pictures:

Mercury:

Venus:

Earth:

Mars:

Which planet seems most like our Moon? Can you think of a reason for this?

Which planet seems to have the most Earth-like features? Name the features.

Which planet has the most cloud cover?

PART 3: JOVIAN MOONS

Moons of	Name	Year Discovered	Diameter (km)	d_{Moon} / d_{Planet}*	Short Description of Picture
Jupiter					
1					
2					
3					
4					
Saturn					
1					
2					
3					
4					
Uranus					
1					
2					
3					
4					
5					
Neptune					
1					

*Divide the moon's diameter by the planet's diameter (you already have planetary diameters listed in the table in Part 1).

Which moon of Jupiter looks volcanic?

Which moon of Jupiter looks ice-covered?

Which moon of Jupiter is the largest?

Which moon of Saturn has an atmosphere?

Which moon of Saturn looks the most like Earth's Moon?

Which moon of Uranus seems to be the victim of the most impacts?

Which moons are larger than the planet Mercury?

Which moon is largest compared to the size of its planet (that is, the largest d_{Moon}/d_{Planet})?

PART 4: LARGE MOONS

Moon	Diameter (km)	Year Discovered	d_{Moon}/d_{Planet}
Earth's			
Charons			

How do the sizes of Earth's and Pluto's moons compare to those of the jovian moons?

How do the sizes of Earth's and Pluto's moons relative to their planets compare to the sizes of the jovian planets' moons relative to theirs?

PART 5: RINGS

Time 28 Sgr first occulted by rings: ____________________ (1)

Time 28 Sgr first occulted behind planet: ________________ (2)

Time 28 Sgr emerged from behind planet: ________________ (3)

Time 28 Sgr emerged from behind last ring: ______________ (4)

Total Time 28 Sgr was behind planet only (3) − (2) = ______________ minutes = t

Total Time 28 Sgr was occulted by rings (4) − (1) = ______________ minutes = T

Divide T by t. T/t = ________________

Density of Saturn d_{planet} = ______________

Density of ______________ (moon name) d_{moon} = ______________

$$\text{Roche Limit} = 2.5\times\left(\frac{d_{planet}}{d_{moon}}\right)^{1/3} = \underline{\qquad\qquad}$$ (compare this value to T/t)

CONCLUSION

In the space below, write a conclusion for this activity. Briefly explain what you did and what you learned from it.

CHECK YOUR UNDERSTANDING 18: THE SOLAR SYSTEM PLANETS, MOONS, AND RINGS

MULTIPLE-CHOICE QUESTIONS

1. Which is not a property of a terrestrial planet?
   a. close to the Sun
   b. high density
   c. low mass and small size
   d. many moons
2. Which is not a property of a jovian planet?
   a. far from the Sun
   b. high density
   c. large mass and size
   d. rings and many moons
3. Which planet is not terrestrial or jovian?
   a. Venus
   b. Earth
   c. Uranus
   d. Pluto
4. Which planet has the most visible impact craters?
   a. Mercury
   b. Venus
   c. Earth
   d. Mars
5. Which planet has the lowest density?
   a. Venus
   b. Jupiter
   c. Saturn
   d. Pluto
6. Which planet was first discovered to have moons?
   a. Jupiter
   b. Saturn
   c. Uranus
   d. Neptune

7. Which moon is the largest in the solar system?
   a. Earth's Moon
   b. Ganymede of Jupiter
   c. Titan of Saturn
   d. Charon of Pluto
8. Which moon is largest compared to its planet?
   a. Earth's Moon
   b. Miranda of Uranus
   c. Triton of Neptune
   d. Charon of Pluto
9. Which planet has the most extensive ring system?
   a. Jupiter
   b. Saturn
   c. Uranus
   d. Neptune
10. The Roche Limit is
    a. the farthest a moon can be from a planet.
    b. the closest a ring can be to a planet.
    c. the boundary between a planet's rings system and moons.
    d. [all of the above].

OPEN-ENDED ACTIVITY

Use **Explore/The Solar System** to observe the alignment of the planets on Jan 1, 2000, at midnight. Make a drawing of what you observe.

VOYAGER: SKYGAZER CD-ROM ACTIVITIES

19
Stars and the H-R Diagram

INTRODUCTION

The *Hertzsprung-Russell Diagram* (or *H-R Diagram*) is a plot comparing two important properties of stars: their temperature and brightness. Temperature is plotted in terms of the star's spectral type, and brightness is plotted in terms of the star's absolute magnitude. You can review what these terms mean and the H-R Diagram on pages 10–15 of *Appendix A: Basic Concepts*. Also, see *MasteringAstronomy Tutorial* 17 for more about the H-R Diagram.

PART 1: THE BRIGHTEST STARS

1. Under the **Chart** menu, change your **Sky View** to **Star Atlas Coordinates** and select **Full Sky View—360°**. Under the **Chart** menu, select **Star Symbols/Prominent Color**.
2. Under the **View** menu, select, in turn, each of the 20 **Brightest Stars**. When the star is centered, observe its color. When its **Data Panel** comes up, record the information requested on the table on the RESULTS sheet. You will find the information under the **General** and **Physical** tabs. If you cannot observe the color, try to guess from its Spectral Type.
3. Based on the data on your table, answer the questions on the RESULTS sheet.

PART 2: THE NEAREST STARS

1. Under the View menu, select, in turn, each of the 20 **Nearest Stars** and follow the same procedure as in Part 1.
2. Based on the data on your table, answer the questions on the RESULTS sheet.

PART 3: THE H-R DIAGRAM

1. Use the Absolute Magnitude and Spectral Type columns on the **Brightest Stars** and the **Nearest Stars** tables to plot each of the stars on the H-R Diagram on the RESULTS sheet.

PART 4: STAR TYPES

1. Based on your H-R Diagram, answer the questions on the RESULTS sheet.

RESULTS SHEET 19 Stars and the H-R Diagram

NAME ______________________ **DATE** ______________ **SECTION** ______________

PART 1: THE BRIGHTEST STARS

Star Name	Apparent Magnitude	Absolute Magnitude	Distance (light-years)	Spectral Type	Temp. (°K)	Color

The stars on the above table are for the most part CLOSE | FAR *(circle one)*. Since they are CLOSE | FAR *(circle one)* but look BRIGHT | DIM *(circle one)*, they should all be intrinsically BRIGHT | DIM *(circle one)*.

Does this appear to be true? YES | NO *(circle one)*.

The temperatures of the stars on the above table appear to be HOT | COOL | BOTH HOT AND COOL *(circle one)*. You may want to review page twelve in *Appendix A: Basic Concepts* before answering this question.

PART 2: THE NEAREST STARS

Star Name	Apparent Magnitude	Absolute Magnitude	Distance (light-years)	Spectral Type	Temp. (°K)	Color

The stars on the above table are for the most part BRIGHT | DIM *(circle one)*, because they are all close.

This means that intrinsically they should all be BRIGHT | DIM *(circle one)*.

Does this appear to be true? YES | NO *(circle one)*.

The temperatures of the stars on the above table appear to be HOT | COOL | BOTH HOT AND COOL *(circle one)*.

Which stars are on both tables?

Are they HOT | COOL | IN-BETWEEN *(circle one)* stars?

Are they BRIGHT | DIM | IN-BETWEEN *(circle one)* stars?

PART 3: H-R DIAGRAM

Plot spectral subtypes to the *right* of the type. For instance, G2 is one space to the RIGHT of G0, K7 is in the middle of the space, three to the right of K0. Plot each star and include its name.

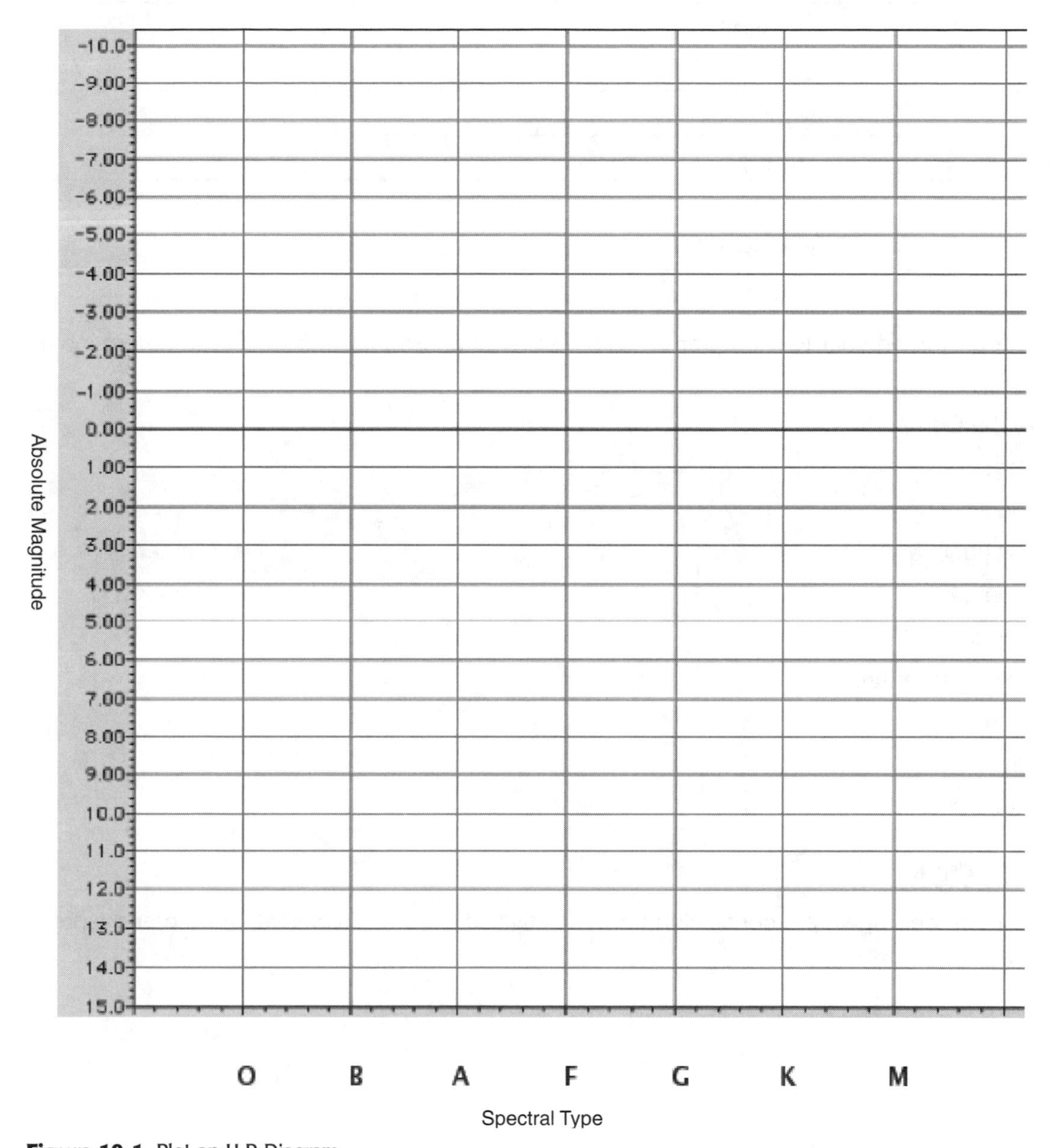

Figure 19-1 Plot an H-R Diagram

PART 4: STAR TYPES

Star types are named in terms of their color and size. The most common colors of stars, in order of increasing temperature, are red, yellow, white, and blue. For the most part, larger stars have more surface area to emit energy, so they are brighter. Dim stars are called *dwarfs*, bright stars are called *giants* and *supergiants*, and those in between can be referred to simply as *main-sequence*. Complete the table on the following page.

Temperature	Brightness	Star Type
cool	dim	red dwarf
cool	bright	
cool	very bright	
hot	dim	
hot	bright	
hot	very bright	
hot	in-between	
cool	in-between	

Use its position on your H-R Diagram (and the stars' **Data Panels,** if you wish) to determine the type of each star.

Alpha Centauri

Vega

Rigel

Betelgeuse

Aldebaran

Regulus

Proxima Centauri

Sirius

Procyon

Tau Ceti

CONCLUSION

In the space below, write a conclusion for this activity. Briefly explain what you did and what you learned from it.

CHECK YOUR UNDERSTANDING 19: STARS AND THE H-R DIAGRAM

MULTIPLE-CHOICE QUESTIONS

1. The brightest stars list is in order of
   a. apparent magnitude.
   b. absolute magnitude.
   c. distance.
   d. spectral type.
2. The nearest stars list is in order of
   a. apparent magnitude.
   b. absolute magnitude.
   c. distance.
   d. spectral type.
3. The H-R Diagram is a plot of
   a. apparent magnitude and distance.
   b. absolute magnitude and distance.
   c. apparent magnitude and spectral type.
   d. absolute magnitude and spectral type.
4. The nearest stars tend to be mostly
   a. bright and hot.
   b. bright and cool.
   c. dim.
   d. hot.
5. Which type is an indicator of temperature?
   a. apparent magnitude
   b. absolute magnitude
   c. spectral type
   d. [none of the above indicates temperature]
6. Which type of star is bright but cool?
   a. blue giant
   b. red giant
   c. red dwarf
   d. white dwarf

7. Where on the H-R Diagram are bright but cool stars?
   a. upper right
   b. upper left
   c. lower right
   d. lower left
8. Which type of star is hot but dim?
   a. blue giant
   b. red giant
   c. red dwarf
   d. white dwarf
9. Where on the H-R Diagram are hot but dim stars?
   a. upper right
   b. upper left
   c. lower right
   d. lower left
10. Which type of star is hottest?
    a. blue giant
    b. red giant
    c. red dwarf
    d. white dwarf

OPEN-ENDED ACTIVITY

Use **Explore/The Solar Neighborhood** and see how many stars you find from both **The Brightest Stars** list you compiled in Part 1 and **The Nearest Stars** list you compiled in Part 2. Make a list of the stars you find. From which list do you expect to find more stars? Explain the reason for your answer. Did your answer turn out to be correct?

20
Galaxies—The Milky Way

INTRODUCTION

Nebulae, *clusters,* and *galaxies* are collectively known as *deep sky objects*. In this activity you will make observations about these objects that will help you understand their nature and the nature of our own Milky Way and its place in the universe.

PART 1: MESSIER OBJECTS

File | Open Settings | Basics | Wide Field Milky Way

The *Milky Way* is a bright band of light that stretches across our entire sky. It is believed to be the light from a majority of the stars in a disk-shaped galaxy of several hundred billion stars of which our Sun is a part. When we look into the Milky Way, we are looking into the disk of the galaxy; when we look elsewhere, we are looking away from it. Your Sky Chart is a view of the entire sky, centered on the Milky Way.

1. Under the **Display** menu, select **Deep Sky Objects**. See that there are basically three types of **Deep Sky Objects: Nebulae, Clusters**, and **Galaxies**.
2. Now, under the **View** menu select the **Messier Catalog**. This is a catalog of **Deep Sky Objects** compiled by the French astronomer Charles Messier. Alternately select the Messier objects (**M-1, M-13, M-31, M-42, M-44,** and **M-57**), click on **Center in Chart**, and record the information requested on the table on the RESULTS sheet. Make sure to examine the pictures of each object, noting both the differences and the similarities between the objects.

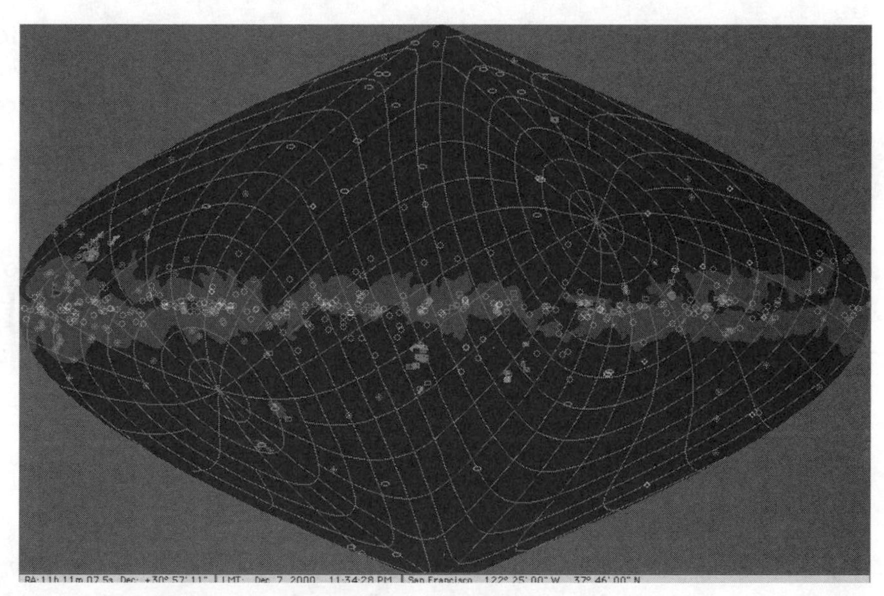

Figure 20-1 Sky Chart centering on the Milky Way

3. Match the pictures with the objects at the end of the RESULTS sheet.

PART 2: NEBULAE

1. Now examine your Sky Chart. Under the **Display** menu, deselect the following **Deep Sky Objects**: **Open Clusters, Globular Clusters**, and **Galaxies** (also deselect **Show Messier Objects**). Leave only the various types of nebulae selected.
2. Now bring up the **Display Panel** (Figure 20-2) from under the **Control** menu and click on the **deep sky object magnitude** button to increase the number of nebulae you can see. Continue clicking until no more new nebulae appear.

   Deep Sky Object Magnitude Button 

Figure 20-2 The **Display Panel** and the deep sky object magnitude button.

3. Are the nebulae evenly distributed through the whole sky or does there seem to be a concentration of them in a certain area? Use the bottom scroll bar or left and right arrow keys to center your Sky Chart on 0° galactic longitude. On the RESULTS sheet draw an approximation of the concentration of the nebulae on the map.

PART 3: CLUSTERS

1. Now change the **Magnitude Limits** under the **Chart** menu to **Small Town—4.5**. This time deselect *all* the various nebulae under **Display | Deep Sky Objects** and select both **Open Clusters** and **Globular Clusters**.
2. Alternate between selecting the **Open Clusters** and **Globular Clusters** and observe where they are concentrated. Use the bottom scroll bar or left and right arrow keys to center your Sky Chart on 0° galactic longitude. On the RESULTS sheet draw an approximation of the concentration of first the open clusters and then the globular clusters on the maps.

PART 4: DISTANCES TO CLUSTERS

1. Click on a few of each of the open clusters and globular clusters to bring up their **Data Panels** and see which are farther away. You can find their **Distance** under the **Physical** tab.
2. Do this as well for some of the various types of nebulae. (All **Deep Sky Objects** can be selected and deselected under the **Display** menu.)
3. Answer the questions on the RESULTS sheet.

PART 5: THE HALO

Since they are closer and seem to lie mostly in the plane of the Milky Way, the nebulae and open clusters are considered part of the galactic disk. Since the globular clusters are farther and not concentrated in the plane, they are thought to form a *halo* around the disk. For this activity, assume that both of these assumptions are true.

Mark an X on your map of the distribution of globular clusters from Part 3 on the RESULTS sheet where you think the center of the galaxy should be. Does the shape of the Milky Way support your choice of this position in any way?

PART 6: OTHER GALAXIES

There was once a debate about whether the deep sky objects we now call galaxies were closer objects within our own galaxy or farther objects very much like our own galaxy.

1. Select **Display | Deep Sky Objects | Galaxies** and deselect all other objects. Click on the **deep sky object magnitude** button (as you did in Part 2) to increase the number of galaxies you can see. Continue clicking until no new galaxies appear. Does the distribution of galaxies (as compared to other deep sky objects) suggest that they are part of our galaxy?

2. Click on some galaxies to bring up their **Data Panels**. Compare their **Distances** to those of some of the globular clusters. What does this tell you?

RESULTS SHEET 20 Galaxies—The Milky Way

NAME ______________________ **DATE** ______________ **SECTION** ______________________

PART 1: MESSIER OBJECTS

Record the information for each *Messier Object* from its **Data Panel**.

Messier Object	Object Name	Object Type	Description of Object Type*	Distance (light-years)
M-1				
M-13				
M-31				
M-42				
M-44				
M-57				

*Use the Tables and Glossary under the Help menu or another source if you need help with this.

PART 2: NEBULAE

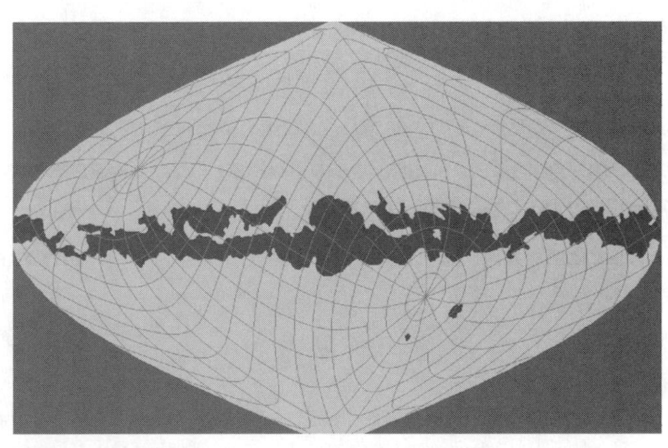

Figure 20-3 Draw the concentration of nebulae on your Sky Chart

PART 3: CLUSTERS

Concentration of Open Clusters

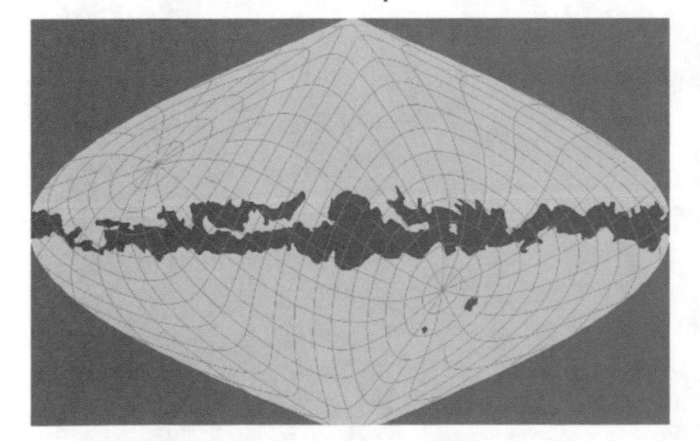

Concentration of Globular Clusters

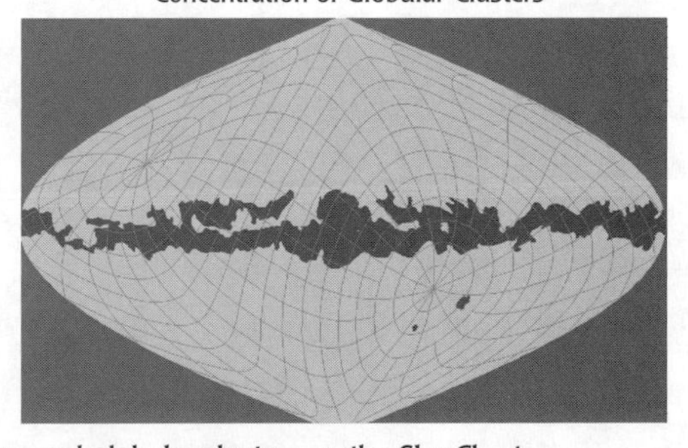

Figure 20-4 Draw the concentration of open clusters and globular clusters on the Sky Charts

PART 4: DISTANCES TO CLUSTERS

In general, **OPEN | GLOBULAR** *(circle one)* clusters seem to be farther away.

In general, nebulae seem to be **FARTHER | CLOSER | ABOUT THE SAME DISTANCE** *(circle one)* as open clusters, and **FARTHER | CLOSER | ABOUT THE SAME DISTANCE AS** *(circle one)* globular clusters.

In general, **NEBULAE | OPEN CLUSTERS | GLOBULAR CLUSTERS** *(circle one)* seem to be the farthest away.

PART 5: THE HALO

Mark on the RESULTS sheet your choice for the position of the center of the galaxy based on the distribution of the globular clusters in the map in Part 3.

Does the shape of the Milky Way support your choice? Explain your answer.

PART 6: OTHER GALAXIES

How does the distribution of galaxies throughout the sky compare to that of the other deep sky objects you looked at?

Does this distribution suggest that they are objects within our galaxies like the nebulae and open clusters?

In general, galaxies seem to be **CLOSER THAN | FARTHER AWAY THAN | ABOUT THE SAME DISTANCE AWAY** *(circle one)* as globular clusters.

This suggests that they are **PART OF OUR GALAXY | VERY FAR AWAY GALAXIES MUCH LIKE OUR OWN** *(circle one)*.

MATCH THE MESSIER OBJECT!

In Figure 20-5 are pictures of the six objects you examined in Part 1. Match each picture with its Messier Catalog number.

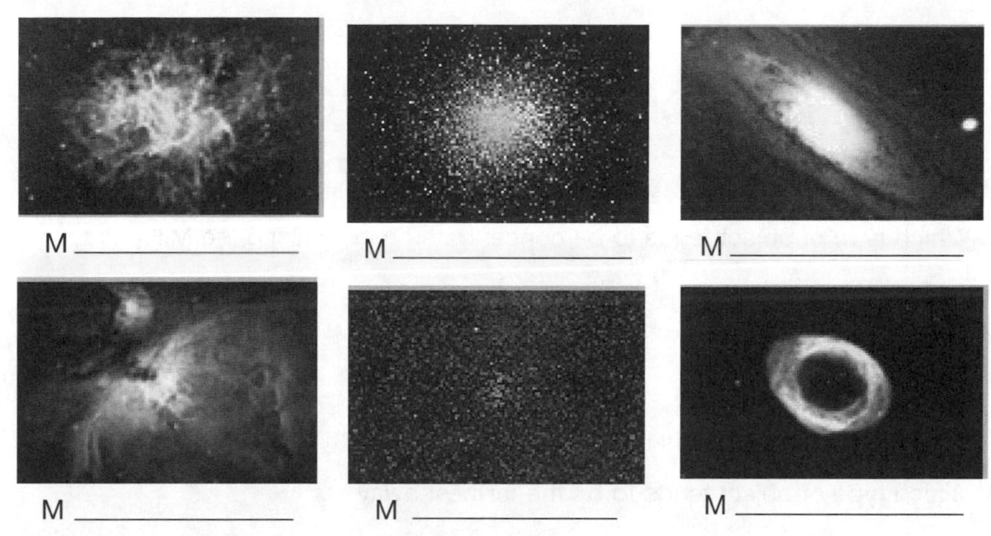

Figure 20-5 Label each object with its Messier number

CONCLUSION

In the space below, write a conclusion for this activity. Briefly explain what you did and what you learned from it.

CHECK YOUR UNDERSTANDING 20: GALAXIES—THE MILKY WAY

MULTIPLE-CHOICE QUESTIONS

1. Which type of object could NOT be a Messier Object?
   a. star
   b. nebulae
   c. cluster
   d. galaxy
2. Which type of object tends to be concentrated mostly along the Milky Way?
   a. nebulae
   b. open clusters
   c. globular clusters
   d. [more than one of the above]
3. Which type of object tends to be the farthest away?
   a. nebulae
   b. open clusters
   c. globular clusters
   d. other galaxies
4. Which type of objects forms a halo around our galaxy?
   a. nebulae
   b. open clusters
   c. globular clusters
   d. other galaxies
5. Which Messier Object is another galaxy?
   a. M-1
   b. M-13
   c. M-31
   d. M-57
6. Which Messier Object is an open cluster?
   a. M-1
   b. M-13
   c. M-44
   d. M-57

7. Which Messier Object is a globular cluster?

   a. M-1

   b. M-13

   c. M-44

   d. M-57

8. Which Messier Object is NOT a nebula?

   a. M-31

   b. M-42

   c. M-44

   d. M-57

9. In which Messier Object can individual stars be clearly made out?

   a. M-1

   b. M-13

   c. M-44

   d. M-57

10. Which Messier Object is the farthest away?

    a. M-1

    b. M-13

    c. M-31

    d. M-57

OPEN-ENDED ACTIVITY

Use the skills you learned in this activity to determine which Messier Objects are visible from your location tonight. Experiment with different **Magnitude Limits** and see if that makes any difference. Record the names and types of objects you may be able to see.

21
The Universe—Hubble's Law

INTRODUCTION

Hubble's Law (or *The Hubble Law*) is a relationship between the distance to a galaxy and the speed it is moving away from us, which was first formulated by Edwin Hubble. In this activity, you will use Hubble's Law and data from real galaxies to determine the *Hubble Constant*, which will allow you to estimate the distance to remote galaxies and even the age of the universe itself. See *MasteringAstronomy Tutorial 21* for more about Hubble's Law.

PART 1: DISTANT GALAXIES

1. Under the **Chart** menu, change your **Sky View** to **Star Atlas Coordinates** and select **Full Sky View-360°**.
2. Under the **Control** menu, bring up the **Display Panel**. Deselect everything that is highlighted except the **Colors** and the **Equatorial Grid** buttons, until your Sky Chart is blank. Then select the **Deep Sky Objects** button. Your **Display Panel** and Sky Chart should look similar to Figure 21-1.

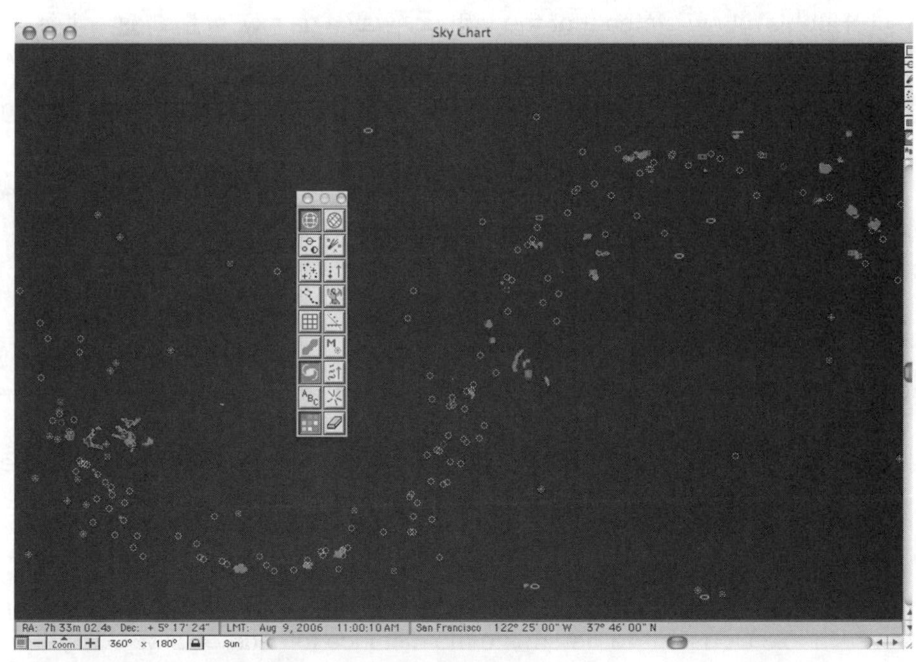

Figure 21-1 Sky Chart showing Deep Sky Objects

3. Now select **Deep Sky Objects** under the **Display** menu and deselect everything except **Galaxies**. Then change the **Magnitude Limits** under the **Chart** menu to **Manhattan—2.5** and nothing should be left except four galaxies. Bring up the **Data Panel** and record the **Object Name, Radial Velocity,** and **Distance** of each of these objects on the table on the RESULTS sheet. When you click on an object, make sure that the **Data Panel** says that it is a galaxy and *not* a constellation.

4. The objects you found are the four closest galaxies to our Milky Way. The Large and Small Magellanic Clouds are satellite galaxies of the Milky Way, and M-31 and M-33 are other members of our Local Cluster. Note the signs of the radial velocities. A positive *Radial Velocity* means that the galaxy is moving away from our Milky Way, a negative *Radial Velocity* means that the galaxy is moving toward our Milky Way.* Mark the appropriate column in the table for galaxies moving toward or away from us.
5. Change **Magnitude Limits** under the **Chart** menu to **Large City—3.5** and record the information for the new galaxy that appears. Record the information in the last row of the table on the RESULTS sheet.

PART 2: MORE DISTANT GALAXIES

1. Change **Magnitude Limits** under the **Chart** menu to **Small Town—4.5** and record the information for the new galaxies that appear.
2. Answer the questions below the table on the RESULTS sheet.

PART 3: VERY DISTANT GALAXIES

1. Change **Magnitude Limits** under the **Chart** menu to **Country—5.5** and record the information for 15 of the new galaxies that appear. Try *not* to record a galaxy that you already did in Part 2.
2. Answer the questions below the table on the RESULTS sheet.

PART 4: HUBBLE'S LAW

1. Hubble's Law states that the farther away a galaxy is, the faster it is moving away from us. Plot about 20 galaxies from your tables in Parts 3 and 4 on the axes on the RESULTS sheet.
2. Draw a line that best fits the points you plotted. "Best fit" means going through as many points as possible with similar numbers of points above and below the line.
3. Determine the slope of your best-fit line. Take the total *rise* of your line (how many km/s the line goes up) and divide by the total *run* of your line (how many mpc it goes to the right). This will be your value for the Hubble Constant.

PART 5: HUBBLE'S CONSTANT

Once you have a value for the Hubble Constant, you can use it to determine the distance to galaxies for which you know only the radial velocity. Find NGC 1365 and NGC 4697 on your Sky Chart. Bring up their Data Panels and notice that their distances are *not* given. You can calculate them with your value of the Hubble Constant:

$$D = V/H$$

Record each galaxy's radial velocity on the table on the RESULTS sheet, then divide the radial velocity by your Hubble Constant to determine the distance to each galaxy.

PART 6: THE AGE OF THE UNIVERSE

The Hubble Law is a relationship between how far away the galaxies are (distance) and how fast they are moving away from each other (velocity). Velocity and distance are related by time, $V = D/T$, and the Hubble

---

*These Radial Velocities are measured with the Doppler effect. See *Mastering Astronomy Tutorial 9* for more about the Doppler effect.
**Note:* If you have graphing software available, it may be preferable to plotting the graph by hand.

Constant is related to distance and velocity by H = V/D. Since V/D = 1/T, then H = 1/T or T = 1/H. The amount of time that the galaxies have been moving away from each other is the reciprocal of the Hubble Constant. The Hubble Constant allows us to calculate the age of the universe. Use the formula on the RESULTS sheet and your value of the Hubble Constant to determine the age of the universe.

RESULTS SHEET 21 The Universe—Hubble's Law

NAME ______________________ **DATE** ____________ **SECTION** ______________________

PART 1: DISTANT GALAXIES

Object Name	Radial Velocity (km/s)	Moving toward us?	Moving away from us?	Distance (kiloparsecs)	Distance (light-years)
M–31					
M–33					
SMC					
LMC					

PART 2: MORE DISTANT GALAXIES

Object Name	Radial Velocity (km/s)	Moving toward us?	Moving away from us?	Distance (kiloparsecs)	Distance (light-years)

What is different about M–49 from all the other galaxies?

All the galaxies are **MOVING TOWARD** | **AWAY FROM** *(circle one)* us.

PART 3: VERY DISTANT GALAXIES

Object Name	Radial Velocity (km/s)	Moving toward us?	Moving away from us?	Distance (kiloparsecs)	Distance (light-years)

The greater the distance to a galaxy, the **GREATER | LESS** *(circle one)* the radial velocity.

PART 4: HUBBLE'S LAW

Rise of your graph: __________ km/s

Run of your graph: __________ mpc

Rise / Run = H = __________ km/s/mpc = your Hubble Constant

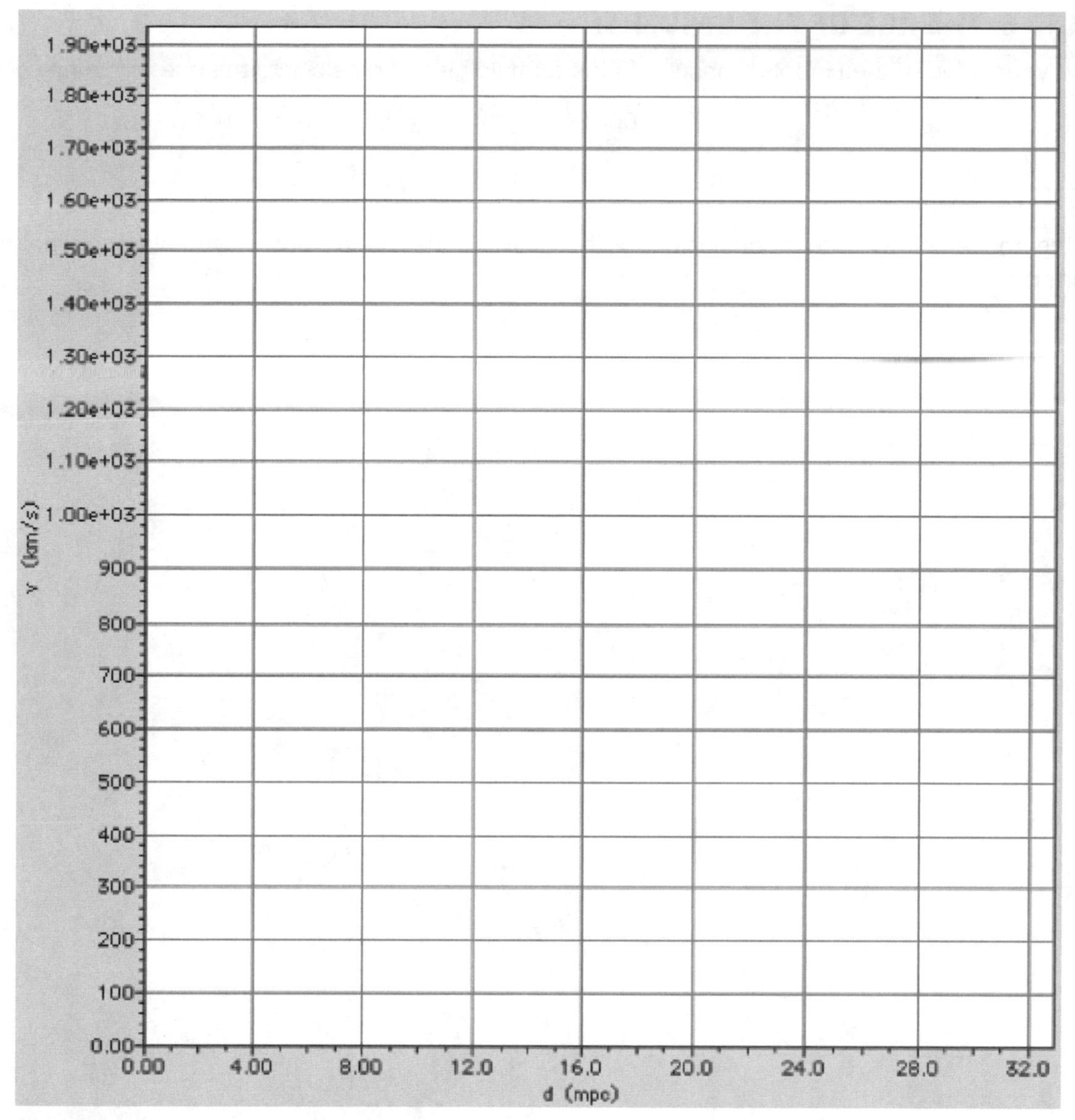

Figure 21-2 Plot a Hubble's Law graph

PART 5: HUBBLE'S CONSTANT

Galaxy	Radial Velocity (km/s)	Hubble Constant (km/s/mpc)	D = V/H (mpc)
NGC 1365			
NGC 4697			

PART 6: THE AGE OF THE UNIVERSE

Use your value of the Hubble Constant in the formula below to calculate the age of the universe.

$T = 1000/H =$ ________ billion years

CONCLUSION

In the space below, write a conclusion for this activity. Briefly explain what you did and what you learned from it.

CHECK YOUR UNDERSTANDING 21: THE UNIVERSE—HUBBLE'S LAW

MULTIPLE-CHOICE QUESTIONS

1. Which object is closest?
   a. M–31
   b. M–33
   c. Small Magellanic Cloud
   d. Large Magellanic Cloud
2. The light from which of the following is blueshifted?
   a. M–31
   b. M–33
   c. Small Magellanic Cloud
   d. Large Magellanic Cloud
3. Which object is moving toward us?
   a. M–31
   b. M–33
   c. Small Magellanic Cloud
   d. Large Magellanic Cloud
4. The light from most distant galaxies is
   a. redshifted.
   b. blueshifted.
   c. [there is about the same number of redshifted and blueshifted galaxies].
   d. [the light from most distant galaxies DOES NOT show any shift].
5. A blueshift suggests that a galaxy is moving
   a. away from us.
   b. toward us.
   c. either toward or away from us depending on how far away it is.
   d. [blueshifts do not give information about a galaxy's motion].
6. A galaxy that is moving away from us will show a
   a. blueshift.
   b. redshift.
   c. either a blueshift or a redshift, depending on how fast it is moving.
   d. [these shifts do not depend on a galaxy's motion].

7. What two properties of a galaxy are compared in Hubble's Law?
   a. mass and distance
   b. mass and recessional velocity
   c. recessional velocity and distance
   d. age and mass
8. What can be determined from a Hubble's Law graph?
   a. the distance to a faraway galaxy
   b. the recessional velocity of a distant galaxy
   c. the Hubble Constant
   d. [all of the above]
9. What information can be determined from the Hubble Constant?
   a. the mass of a galaxy
   b. the age of a galaxy
   c. the mass of the universe
   d. the age of the universe
10. The Hubble Constant is the ________________ of a Hubble's Law graph.
   a. rise
   b. run
   c. slope
   d. [none of the above]

OPEN-ENDED ACTIVITY

Try using the radial velocity and distance of several individual galaxies and calculate a Hubble Constant for each one of them. Record the values of H for each galaxy. How do the individual values compare to the value you got from your graph? Do you think that the individual values are more or less accurate than the value from your graph? Explain your answer.

Appendix
BASIC CONCEPTS

Basic Concepts

Voyager III allows you to recreate events that cannot be shown in a conventional planetarium or experienced in a lifetime of observing the sky. Before using the software, you should have a good understanding of the astronomical concepts and terms discussed here. This brief overview is intended for those with little background in astronomy, so feel free to skip over any topics with which you are familar.

The Celestial Sphere

Ancient astronomers perceived the sky as a large sphere with the Earth at its center. They thought the stars were attached to the surface of this great sphere, and as it rotated once each day, the stars would rise and set as they were carried across the sky. We know today that this sphere is not real. The stars and planets are at great distances from the Earth, and their apparent daily motion across the sky results from the Earth's spinning on its axis.

But this image of a celestial sphere which surrounds the Earth at some great distance remains a useful concept. Astronomers have created a grid of reference lines and points on the celestial sphere to describe the position of each star, planet, and galaxy. Every object has a numerical address in the sky and that address is on the celestial sphere .

Equatorial Coordinates

We use a coordinate system of longitude and latitude to locate any point on the surface of the Earth. On the celestial sphere we use a similar system called equatorial coordinates, which are based on the Earth's poles and equator.

The Earth's axis of rotation, extended outward from the North Pole, intersects the celestial sphere at a point called the North Celestial Pole. The star closest to this point, Polaris, is often called the North Star. A similar extension from the South Pole marks the South Celestial Pole. The Celestial Equator is the projection of the Earth's equator onto the celestial sphere.

All points along the celestial equator are equidistant from the north and south celestial poles. To define the location of an object on the celestial sphere, we first determine its angular position along the celestial equator. This coordinate is right ascension, and it is analogous to longitude on the Earth. As Greenwich marks the zero of longitude on Earth, the vernal equinox is the zero of right ascension in the sky. This point marks the Sun's position on the first day of spring, when the Sun crosses the celestial equator . The vernal equinox is currently in the constellation of Pisces, but the position changes slowly over time.

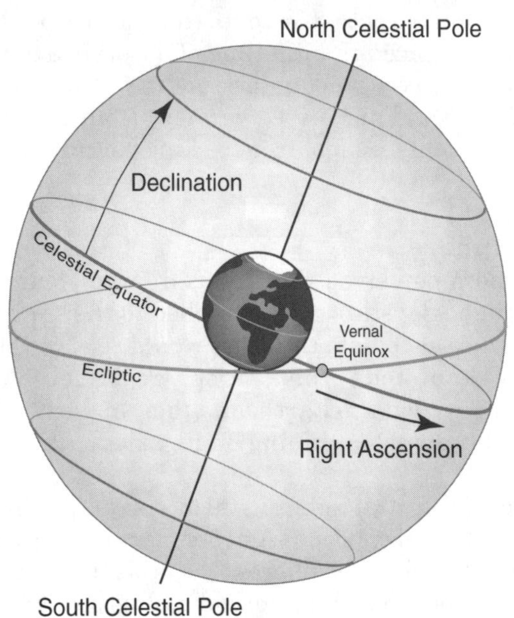

The Celestial Sphere

Right ascension is measured eastward in units of time (hours, minutes, and seconds), starting at 0 hours and continuing to 24 hours. An hour of right ascension is equal to 15 degrees on the celestial equator (1/24 of 360 degrees)

The second equatorial coordinate is declination, and it measures an object's angular distance from the celestial equator. Declination is measured in degrees, from 0° at the celestial equator to 90° at the celestial poles. Objects above the celestial equator have positive declination, and those below have negative declination. The North Celestial Pole is +90° and the South Celestial Pole is -90°. The position of any object in the sky can be defined by the coordinates of right ascension (abbreviated RA) and declination (abbreviated Dec). All celestial objects have an address on the celestial sphere - their RA and Dec. For example, the bright star Sirius is found at RA 6h 44.8m and Dec -16° 42'.

Altazimuth Coordinates

Altazimuth coordinates describe the position of an object with reference to the local horizon. Azimuth is the angular measure along the horizon, beginning at 0° in the north, through 90° in the east, 180° at south, and 270°in the west.

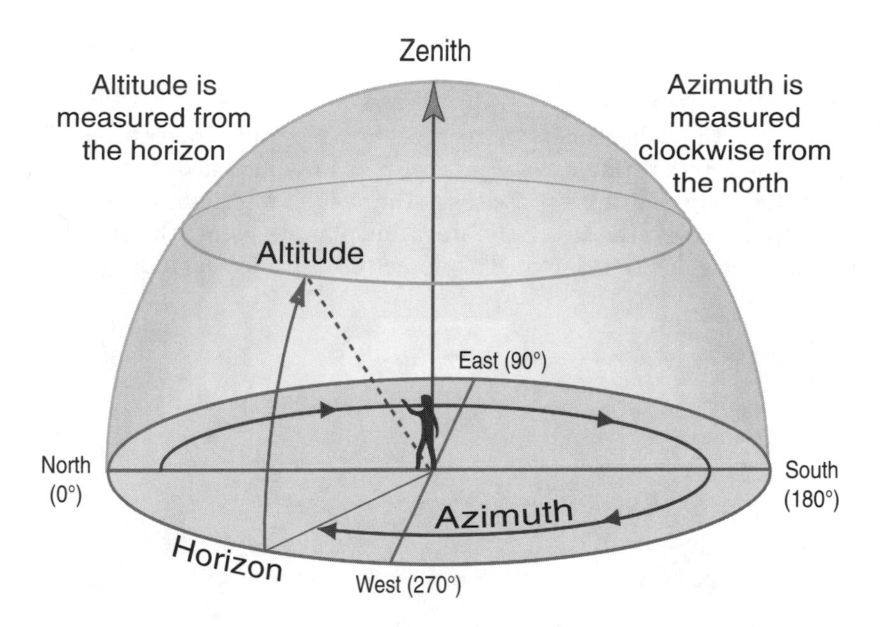

Altazimuth Coordinates

Altitude is measured from 0° at the horizon to 90° at the zenith point directly overhead. A star with an azimuth of 225° and an altitude of +30° would be located 30° above the southwestern horizon. Points below the horizon have negative altitudes. Altitude is sometimes called elevation, but in astronomy, elevation refers to the distance one is above sea level (in feet or meters). The zenith is the point overhead. An object at the zenith will have an altitude of +90°. The nadir is the point opposite the zenith with an altitude of -90°. When you point directly upward, you are pointing toward the zenith. Point downward, and you point toward the nadir.

The meridian is the line that passes through the celestial poles and the zenith. It is the projection of the observer's Earth longitude on to the sky. A transit occurs when an object passes across the meridian. At this time, the object is at its maximum altitude in the sky.

The altazimuth coordinates of an object are local coordinates. They apply at a particular location and at a particular time. At a different location on the Earth, an object will have different altazimuth coordinates. The altitude and azimuth of a star are constantly changing as the Earth rotates, but the star's right ascension and declination on the celestial sphere remain fixed.

View of the Night Sky

From mid-northern latitudes, we see the stars moving across the sky from east to west. As the Earth rotates, the sky appears to pinwheel about the North Celestial Pole. In the early evening a star rises in the east; at midnight it is on the meridian; and by sunrise it sets the west. The stars and planets seem to be attached to a great dome that encircles the Earth. The dome rotates once each day.

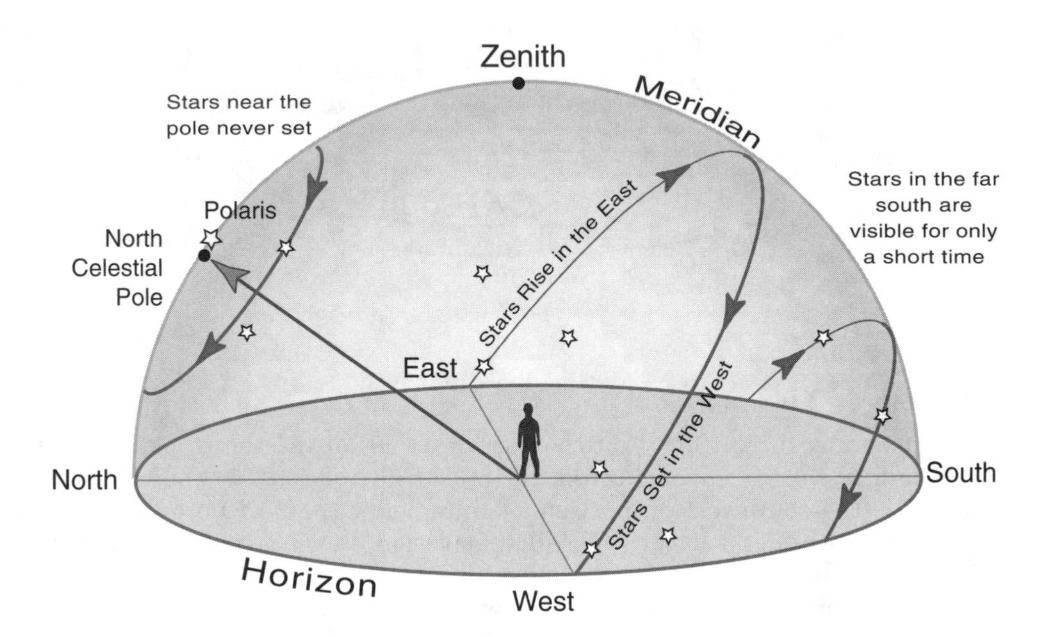

There is a part of the sky that is always visible. Stars near the North Celestial Pole dip near the horizon, but they do not set; these stars are called circumpolar.

One bright star, Polaris, is located less than one degree from the pole. As the sky turns, Polaris, being near the center of rotation, is almost stationary. This makes it a beacon, marking the direction of north for all observers and mariners in the Northern Hemisphere. Polaris is the North Star.

Stars far to the south move on small arcs above the southern horizon. They are visible for only a short time. Stars very far south near the South Celestial Pole do not rise at all, and they remain unseen for observers in the north.

The Ecliptic

As the Earth orbits the Sun, we see the Sun moving against the background stars. The Sun's apparent annual path against the background stars is the ecliptic. The planets are always seen within about 18 degrees of the ecliptic, since they orbit the Sun in nearly the same plane. The ecliptic is the zone of planets, as well as the Sun.

As the Sun moves along the ecliptic, it passes through particular groups of stars. Ancient observers organized these groups into constellations and named them after particular animals. This group of twelve constellations is the zodiac, which derives from the Greek word for "animals". These constellations came to be known as the sign of the zodiac, and they are a center piece of astrology.

The zodiac includes some prominent constellations such as Taurus the bull and Leo the lion. The majority are more obscure and not easy to identify. Astronomers eschew the term "signs of the zodiac" because of its astrological overtones. For astronomers, the constellations along the ecliptic are the "zodiacal constellations". In this special region, we see the complex motions of the planets as they orbit the Sun.
If you observe the sky at sunrise over several weeks, you can see how the Sun moves eastward against the background stars. A constellation that is near the eastern horizon at sunrise will be considerably higher in the sky at sunrise one month later. The Earth' position in its orbit has changed, and from this new perspective we see the Sun against new background stars. This annual cycle

creates the changing pattern of constellations that are visible at night. It is the natural calendar of the sky.

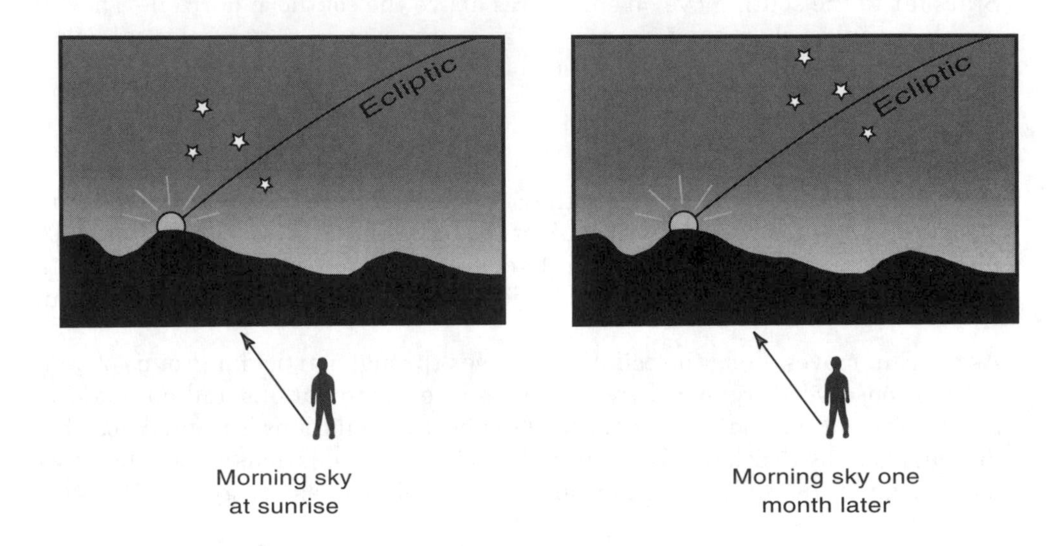

Morning sky at sunrise

Morning sky one month later

Like the Sun, the planet's positions are continually changing with respect to the background stars. The word 'planet', in fact, derives from a Greek word meaning "wanderer". In comparison, the stars are essentially fixed on the celestial sphere. They are in truth moving, but because of great distances their relative positions appear nearly fixed for many centuries.

The positions of the planets are often described using ecliptic coordinates. These are similar to equatorial coordinates, except that they are defined by the plane of the ecliptic, rather than by the Earth's equator. Because of the tilt of the Earth's axis, this coordinate system is inclined 23.5° to the equatorial system. The vernal equinox is the starting point of ecliptic longitude. It is measured in degrees eastward along the ecliptic. Ecliptic latitude is measured in degrees north or south of the ecliptic. The north ecliptic pole is in the constellation Draco.

The Seasons

The Earth has seasons because its axis of rotation is tilted with respect to the plane of its orbit. During the summer months in the Northern Hemisphere, the axis leans towards the Sun resulting in long days and short nights. The first day of summer occurs about June 21 each year; this is the summer solstice. At the same time in the Southern Hemisphere, the axis leans away from the Sun. The winter days are short and the nights are long. June 21 is the winter solstice in the Southern Hemisphere.

In six months the Earth is on the opposite side of its orbit. The axis has remained fixed in space, but from this position in the orbit, the northern tip of the axis now leans away from the Sun. This is the winter solstice. At this same time in the Southern Hemisphere, the axis tilts toward the Sun. Summer begins and the days grow long.

About March 21 the Earth is midway between winter and summer. The Sun lies on the celestial equator, and day and night are of equal length all over the Earth. This is the spring equinox (meaning "equal night") when spring begins in the north and fall begins in the south. Around September 21 is the fall equinox when spring begins in the south and fall begins in the north.

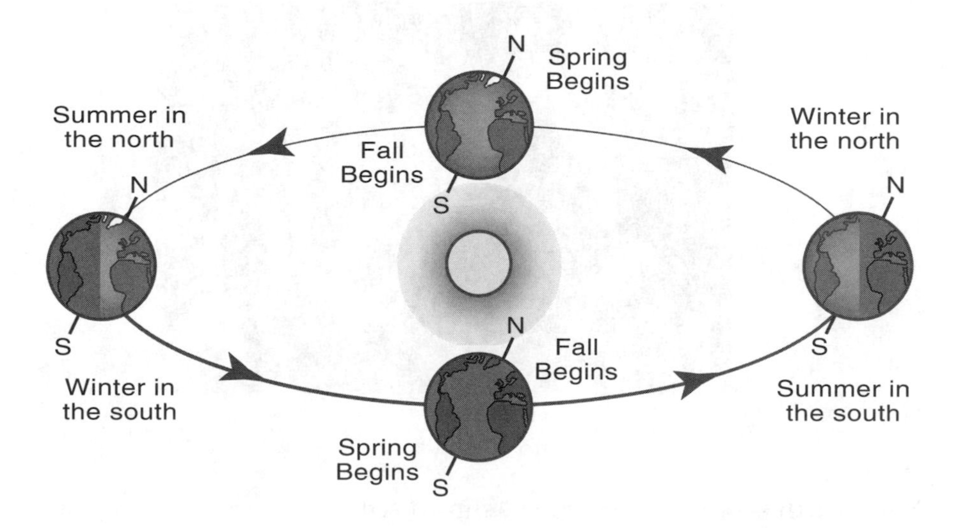

Throughout the year, the Sun appears alternately above and below the celestial equator for six month intervals. When the Sun is above the celestial equator and high in the sky in the Northern Hemisphere, it is low in the sky in the Southern Hemisphere. Seasons alternate between the hemispheres every six months. Summer in Australia and South America is winter in North America and Europe.

Time of Day

The rotation of the Earth is the clock used in everyday life. A day is one rotation of our planet. However, to determine when the Earth has completed one rotation, we must choose some marker in the sky from which we measure. There are several kinds of days depending on the celestial body we choose as our marker.

Our activities are regulated by the cycle of day and night; it is natural to use the Sun as our time keeper. We can define a day as the period between successive passages of the Sun across the meridian. This does not work well, since the length of the day varies continuously throughout the year. Due to the slight eccentricity of the Earth's orbit and the tilt of the Earth's axis, the Sun's eastward motion against the background stars is not uniform. The Sun moves faster in the winter and slower in the summer. The 24 hour day would in some months be 10 minutes less, and in other months 15 minutes more. This is one reason that clocks like the classic sundial provide very crude time keeping.

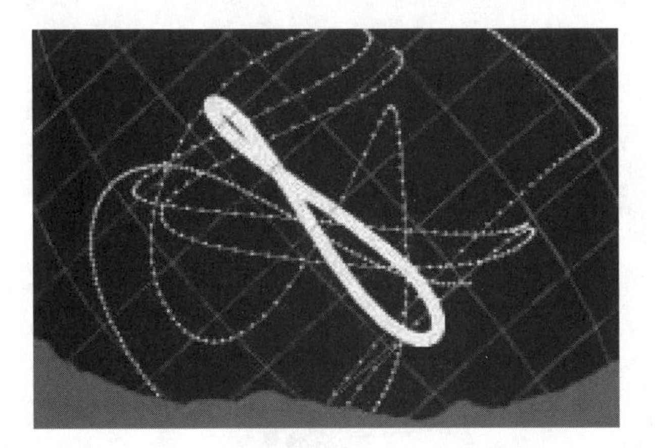

The Analemma

To resolve this difficulty, a fictitious sun is used, one that moves at a uniform rate throughout the year. This is called the mean sun, and it gives us a uniform day. The mean solar day is the average of the apparent solar days over one year. Mean solar time is what we use in our daily lives. The difference between apparent solar time (using the real Sun) and mean solar time is the Equation of Time. Over one year it can be as large as 16 minutes.

The Equation of Time is commonly plotted on globes of the Earth, where it appears as a figure "8" curve. This figure is known as the analemma; it shows the Sun's position at exactly one day intervals over an entire year. The Sun moves north and south across the celestial equator because of the seasons. The Sun shifts in longitude because of the difference between the mean and apparent solar day.

In 1884 an international conference divided the Earth into 25 time zones. The meridian of Greenwich, England was set as the zero of longitude, and the time zones were measured in steps of 15 degrees from this location. The exact boundaries of each zone would vary depending on local politics and circumstance.

The Sidereal Day

A sidereal day is the period of the Earth's rotation in relation to the stars, rather than the Sun. A sidereal day is about 4 minutes shorter that a mean solar day. Over 24 hours, the Earth moves about 1/360 of its orbit. We see the Sun drifting eastward against the stars by 1/360 of a circle, which is 1 degree. In terms of the Earth's rotation, this amounts to 4 minutes in 24 hours. If the Sun and a star are at the same position in the sky, after one rotation of the Earth, the Sun will have drifted eastward, and the star will pass overhead 4 minutes before the Sun. The sidereal day is approximately 1436 minutes, 4 minute less than 24 hours.

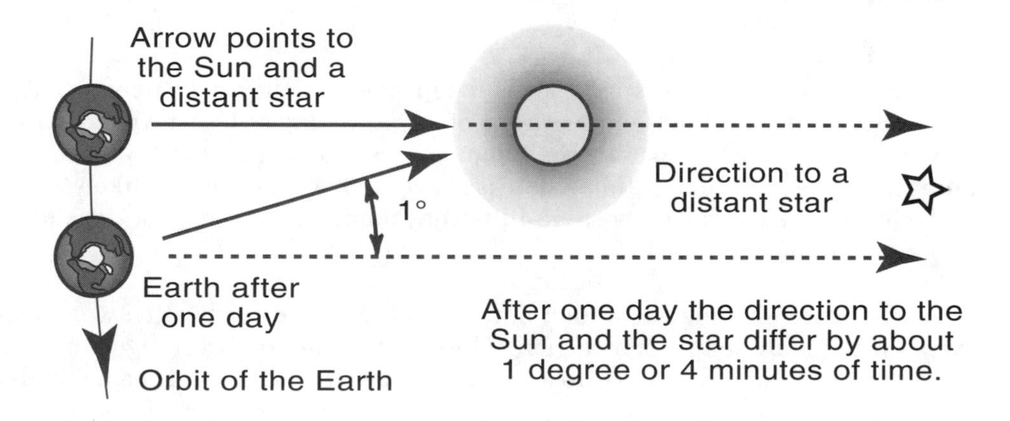

The Sidereal Day

Sidereal time is "star time". It is the elapsed time in hours, minutes, and seconds since the vernal equinox crossed the meridian. The right ascension that is currently on the zenith is the local sidereal time.

Universal and Dynamical Time

Universal Time (UT) is based on Greenwich Mean Time. UT is a 24 hour clock widely used by astronomers and navigators. Universal Time is based on the rotation of the Earth. It includes a correction for the slight wandering of the Earth's geographic pole. The local time in New York is 12 hours behind the time in Singapore, but both cities report the same Universal Time.

The Earth's rotation is slowing down by about 1 second per year, so the average length of the day is increasing. Precise astronomical calculations require a dynamical time scale that is free of the irregularities caused by the Earth's rotation. Dynamical time is based on atomic clocks. Universal time is continuously falling behind dynamical time. In the year 1990, the difference was about 57 seconds. In 2010, UT will lag behind dynamical time by 67 seconds.

Astronomcal events that occurred several thousand year ago, such as an ancient eclipse, can be calculated with good precision. The date of the event is in terms of dynamical time. Knowing the UT time of the event requires careful estimates of how the Earth's rotation has slowed over the centuries. These estimates lead to time corrections of several hours for events in Roman times, and larger corrections of more than a day for earlier epochs.

Julian Date

Astronomers use a date system that is independent of the civil calendar. The Julian date is a continuous enumeration of days starting at Greenwich noon on January 1, 4713 B.C. A particular Julian date is referred to as the Julian day number. Since the starting Julian date is nearly 7000 years ago, any likely event under discussion will have a positive Julian day number. January 1 of 2000 is the Julian day number 2452545.0.

One advantage of using the Julian date is that the time interval between two events is found by simple subtraction. Moreover, there are direct algorithms to convert any Julian day number to a calendar date. If you reckon time using the Julian date, there is no Y2K problem. January 1, 2000 has no special designation or significance.

Star Magnitudes

The brightness of the stars was first estimated by the Greek astronomer Hipparchus in the second century B.C. Hipparchus' system divided the stars into five magnitude classes. The brightest stars were first magnitude, and the faintest stars were fifth magnitude, with all others falling somewhere in between.

This system has been formalized and extended in modern times to apply to all celestial bodies. In general, the brighter an object, the smaller the number that represents its magnitude. The brightest objects, like the Sun or the Moon, actually have negative magnitudes.

Precise measurements show that the brightest star in the night sky is Sirius, which shines at magnitude -1.46. The planet Venus is even brighter at magnitude -4.6, and the daytime Sun has magnitude -27. The faintest objects visible to the unaided eye are about magnitude 6.0. Using a small telescope, stars and galaxies of 10th magnitude are readily seen. The world's largest telescopes are capable of detecting objects of magnitude 25, and the Hubble Space Telescope reaches magnitude 28 and fainter.

The magnitude scale is not linear. A difference of one magnitude corresponds to brightness ratio of about 2.5. Therefore, a second magnitude star is 2.5 times brighter than a third magnitude star, and it is more than 6 times brighter (2.5x2.5) than a fourth magnitude star. The magnitude scale is precisely defined such that a magnitude difference of 5.0 represents a brightness ratio of 100. A difference of one magnitude corresponds to the fifth root of 100 (2.512).

Measuring the apparent magnitude of an object involves several factors: the object's distance (objects farther away appear fainter); the wavelengths considered (some objects are brighter at infrared or ultraviolet wavelengths than in visible light); the sensitivity of the detector (the eye is less sensitive to blue light than standard photographic film). Voyager III uses either visual or photographic magnitudes taken from various astronomical catalogues.

Astronomers also use a quantity called absolute magnitude. It specifies the magnitude of a star if it were seen at a standard distance of 10 parsecs (32.6 light years). With absolute magnitude, one can compare the "true" brightness of the stars, since the comparison is always at the same distance.

Star Spectra

It is apparent to any observer that the stars vary in color as well as brightness. The color is an indication of the star's temperature and spectra. The blue stars are hot

(20,000-30,000 °K), and the red stars are cool (2,000-3,500 °K). The spectrum of a star usually has a sequence of dark lines marking wavelengths that are absorbed by the chemical elements in the star's atmosphere. The temperatures of the stars vary greatly, which strongly effects the strengths of their spectral lines.

Early this century astronomers divided stellar spectra in a sequence of classes designated by letters. This system has evolved into the present-day set of spectral classes: O, B, A, F, G, K, and M. A few stars do not fit into this scheme, and they have been given their own special classes: R, N, S, W. Modern instruments are able to resolve a star's spectra into tenths of a class. For example, you will see stars of spectral type B9, G4, and K3.

Principle Spectral Types

type	color	temperature	description	example
O	blue	25,000 to 40,000	strong lines of ionized helium and highly ionized metals	Zeta Orionis
B	blue	11,000 to 25,000	lines of neutral helium, weak hydrogen lines	Spica Rigel
A	blue to white	7,500 to 11,000	strong lines of hydrogen, ionzed metals, weak helium lines	Vega
F	white	6,000 to 7,500	hydrogen lines weaker than strong line of neutral metals	Canopus
G	white to yellow	5,000 to 6,000	hydrogen lines weaker than type F, strong lines of calcium and other neutral metals	Sun Capella
K	orange to red	3,500 to 5,000	numerous lines of neutral metals	Arcturus
M	red	3,000 to 3,500	numerous lines of neutral metals, strong molecular bands	Antares

The O and B stars are hot and blue. The bright star Sirius is of spectral type A. The Sun is a yellow G type star. The K and M stars are red in color and comparatively cool. The brightest of these are the red giants such as the stars Antares and Betelgeuse.

Within a given spectral class, stars are divided into luminosity classes. These classes permit differentiating, for example, between a red giant and red dwarf. The luminosity classes are:

I supergiant
II bright giant
II giant
IV subgiant
V main-sequence dwarf
VI subdwarf
VII white dwarf

Hertzsprung-Russell Diagram (HR)

There is wide variation in the luminosities, spectral types, and temperatures of the stars. One of the great scientific discoveries of this century is understanding the connection between the luminosity and temperature in the stars. Over a star's lifetime these properties change as it evolves to a final fate determined by its mass.

In the first decade of the 20th century, the Danish astronomer Hertzsprung and American astronomer Russell made the first study comparing the brightness (absolute magnitude) to the spectral type of the stars. In 1914 they published what has come to be known as the Hertzsprung-Russell (HR) diagram. For nearly a century the HR diagram has been a powerful visual tool in understanding the properties and evolution of stars.

The HR diagram is a graph with the absolute magnitude on the vertical axis and the spectral type on the horizontal axis. The spectral sequence is equivalent to a temperature sequence, with the hot O and B stars on the left, and the cool K and M stars on the right.

For most stars, there is direct relationship between spectral type and luminosity. The HR diagram features a broad diagonal band known as the main sequence. It shows a direct connection between the luminosity and the spectral type or temperature. The most luminous stars are very hot with spectral types of O, B and A; examples are Rigel in Orion and Murzim in Canis Major. These stars are in the upper left of the HR diagram.

Average stars like the Sun and Alpha Centauri are of spectral type G in the middle of the graph. The dim K and M stars, such as Barnard's Star and Kruger 60, are cool and red at the bottom right of the main sequence. In general, stars have increasing luminosity with increasing temperature.

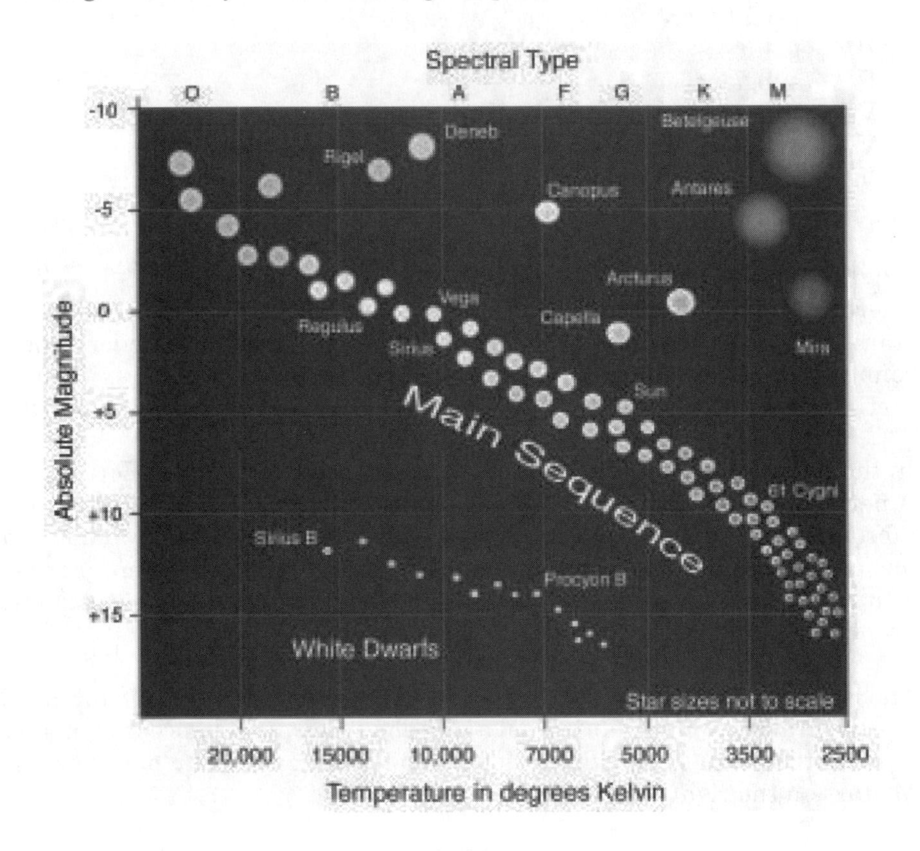

The Hertzsprung-Russell Diagram

What determines where a star lies on the main sequence ? Modern astronomy has revealed that the initial mass of a star is the key to understanding a star's evolution. The masses of stars are found from binary star orbits; using the orbital period and the distance to a binary star system, the mass of the individual stars can be calculated. The range of stellar masses is .01 to 1000 times that of the Sun.

A star's mass is its fuel reservoir. The most massive stars burn hydrogen at a prodigious rate. They are brilliant, but their lives are short - no more than a few million years. Lower mass stars like the Sun are far less luminous, but they can shine for billions of years.

Red Giants and White Dwarfs

There are many stars not on the main sequence. In the upper right corner of the HR diagram is a group known as red giants. Some of the more famous stars in the sky are red giants: Betelgeuse in Orion, Antares in Scorpius, Mu Cephei, and Aldebaran in Taurus.

Toward the end of a star's life, internal changes in its structure produce an increase in luminosity causing it to swell in size. The surface cools as it approaches 100 times its previous diameter. The star becomes a red giant.

The main sequence is the region of stability in the HR diagram. Stars live out the majority of their lives with a balance between gravitation and the radiation pressure from the interior. When this balance ends, they evolve to become red giants. All red giant stars were once on the main sequence.

In the lower left corner of the HR diagram is a unique group of low luminosity stars known as white dwarfs. This is the stellar graveyard. A relatively low mass star like the Sun will eventually become a white dwarf.

After spending 90% of its life as a main sequence star, a sun-like star evolves to a red giant. At the final stages, a red giant expels its outer envelope of material, and the core collapses to the size of the Earth. Depleted of nuclear fuel, its luminosity comes from the residual heat in the core; but the now revealed core is very hot giving the white dwarf its high surface temperature. The most famous of the white dwarfs is the companion of Sirius. It is visible in a modest size telescope. The bright star Procyon also has a white dwarf companion.

Stars more massive than 1.4 solar masses are predicted to collapse to the superdense state of a neutron star. Individual atoms are broken down forming a body composed of only neutrons. The star now measures typically 20 kilometers– in diameter, but with the density of an atomic nucleus. Several neutron stars have been identified as rapidly rotating radio sources know as pulsars.